# Thinking outwards 2

Thomas Nelson and Sons Ltd
36 Park Street London W1Y 4DE
PO Box 18123 Nairobi Kenya
Thomas Nelson (Australia) Ltd
597 Little Collins Street Melbourne 3000
Thomas Nelson and Sons (Canada) Ltd
81 Curlew Drive Don Mills Ontario
Thomas Nelson (Nigeria) Ltd
PO Box 336 Apapa Lagos
Thomas Nelson and Sons (South Africa) (Proprietary) Ltd
51 Commissioner Street Johannesburg

First published 1971

0 17 421040 X

Filmset and Printed Offset Litho
by Cox and Wyman Limited
London, Fakenham and Reading

**Understanding new mathematics**

**George W. Rodda** B.Sc.

Principal Lecturer in Mathematics,
Alsager College of Education, Cheshire

# Thinking outwards 2

**Nelson**

# Contents

6   Reflections
8   Regular Polygons
10   Parallels
12   Venn Diagrams
14   Factors and Divisibility
16   Numbers from Tallying
18   Addition and Subtraction of Fractions
20   Routes
22   Binary Numbers
24   Arrow Graphs and Series
26   Constant Change
28   Some Special Factors
30   How to Convert
32   Speeds from Graphs
34   Capacity
36   Averages
38   The Thickness
40   Parts of a £
42   Multiplication and Division of Fractions
44   Solution Sets
46   Binary Arithmetic
48   The Load
50   Polyhedra
52   How to Measure Angles
54   The Sum of the Angles
56   Finite Arithmetic
58   Arrays and Relations

60   Circumference
62   Some Properties of a Cylinder
64   How to Use Pie Charts
66   Cones
68   Routes and Networks
70   Multiplication of Decimals
72   Travel Graphs
74   Volume
76   Decimal Answers
78   Area of Surfaces
80   Curved Graphs
82   Wholesale and Retail
84   Time
86   Division by Decimals
88   Ratio
90   Congruence and Similarity
92   Area of Triangular Regions
94   Numbers from Ordered Pairs
96   Value for Money
98   Gradients
100   The Percentage
102   Solutions from Venn Diagrams
104   Rotations
106   Weight per Cubic Centimetre
108   Measurements from Angles
110   Factorials
112   The Chance
117   Answers

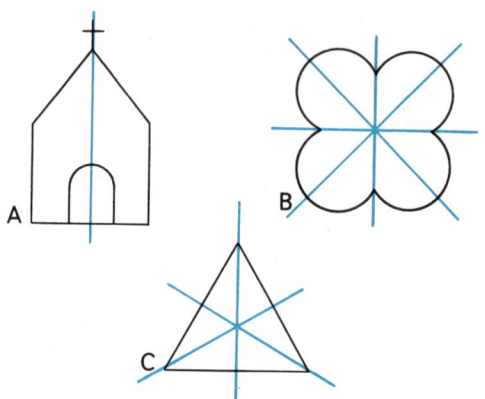

## Let's Discover    Reflections

1. Complete the following sentences:

   a. In picture A the blue line is a line of . . . .

   b. In picture B there are . . . lines of symmetry.

   c. The . . . of symmetry is marked with ✳ in picture C.

   d. Shape A has . . . centre of symmetry.

---

● Shapes which are balanced about a line are **symmetrical** about the line.

---

2. Print the word JAM in pencil on a sheet of paper.
   Draw the line segment *l*.
   Place a piece of tracing paper over the word, with one edge on *l*.

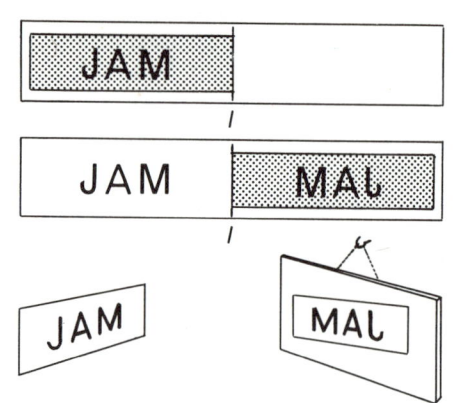

   a. Carefully trace the word and then flip the paper over. Keep the edge on *l*.

   b. Which two letters still look the same?

   c. Hold your paper up to a mirror. Do you see MAL in the mirror?

   d. Is the pattern symmetrical about line *l*?

---

● A **line of symmetry** reflects one half of the picture into the other half.

---

3. Hold a mirror on its edge along this broken line. Do you see the line of symmetry of "8"?
   Letters which have a line of symmetry are not altered when reflected in a mirror which is parallel to the line of symmetry.

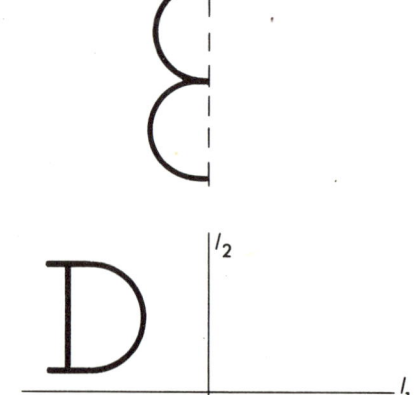

4. a. Draw the image of D in *l₁*, and then in *l₂*.

   b. Which **image** looks like D?

   c. Draw the line of symmetry on D.

1. Here is a word which is unchanged when reflected in $l_1$.

   a. Find another three-letter word which is unchanged when reflected in $l_1$.

   b. Find a three-letter word which is unchanged when reflected in $l_2$.

2. Find a shoe box or some other rectangular box with a loose lid.

   a. How many ways are there of putting on the lid?

   b. How many lines of symmetry has the lid?

   c. Draw the lines of symmetry on the lid.

3. Copy this diagram which shows a square, a rectangle, an equilateral triangle, and a regular pentagon.

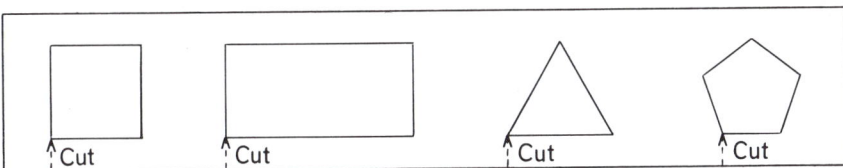

   a. Cut out each shape starting on the dotted line.

   b. Copy and complete the following table showing the number of ways of replacing each shape.

| Shape | Number of ways of putting back without turning the piece over | Total number of ways of putting the piece back | Number of lines of symmetry |
|-------|-----|-----|-----|
| 1<br>2<br>3<br>4 | | | |

   c. In how many ways can a cork be replaced in a bottle?

4. Draw the reflection of each shape in the coloured lines marked. How many lines of symmetry has each completed shape?

   a. 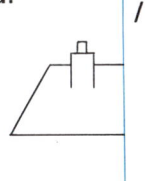 $l$

   b.  $l$

   c. 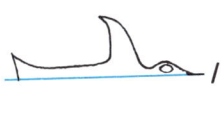 $l$

   d.  $l$

7

The shape of this nut is formed by the union of **line segments**.

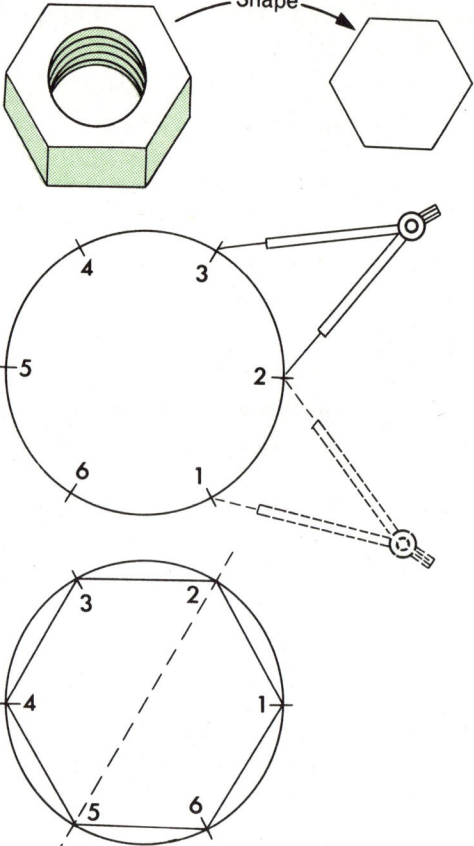

Shape

1. **a.** How many line segments make the **shape**?

   **b.** How many vertices has the shape?

2. Draw a circle with a radius of 5 centimetres (cm).
   Keep your compasses open at 5 cm and step round the circle.
   Six equal steps can be taken round the circumference of the circle.

3. **a.** Join the points 1 to 2, 2 to 3, 3 to 4, . . . to form a six-sided shape.

   **b.** Cut out your six-sided figure and fold it along the dotted line of symmetry.

   **c.** How many more lines of symmetry can be creased?

   **d.** Mark with X a centre of symmetry on your shape.

Here are some more shapes:

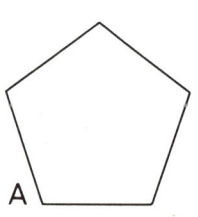
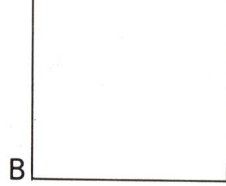
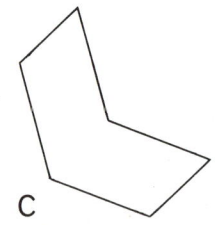
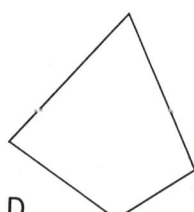

A     B     C     D

4. **a.** Are all the shapes formed by the union of line segments?

   **b.** Are all the shapes simple closed curves?

● A **simple closed curve** formed by the **union of line segments** is called a **polygon**.

**5.** For each of the shapes A, B, C, and D answer these questions:

    **a.** Are all its angles equal?

    **b.** Are all its sides the same length?

● A polygon with all its angles equal and all its sides equal in length is called a **regular polygon**.

This is a regular polygon which is a pentagon.

This is not a regular pentagon.

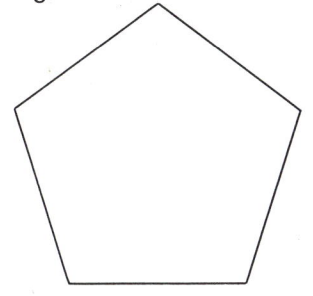

**Let's Try**

**1.** **a.** Complete this table for these shapes:

| Shape | Sides equal in length? | Angles equal? | Regular? |
|-------|------------------------|---------------|----------|
| A | Yes | | |
| B | | | |
| C | | | No |
| D | | | |

    **b.** How many lines of symmetry has each shape?

A

B

C

D

**2.** Here is a regular polygon which is called an equilateral triangle.

    **a.** Make a larger equilateral triangle by tessellating a shape like this.

    **b.** Build another regular polygon from shapes like this.

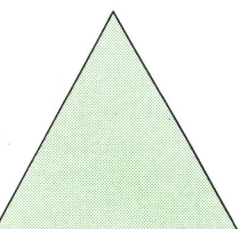

**3.** Make a list of all the regular polygons which you can find in your classroom.

**Parallels**

On a globe you will find lines which are
drawn round it ⟶
and lines which are drawn down it
through the North and South poles ↓

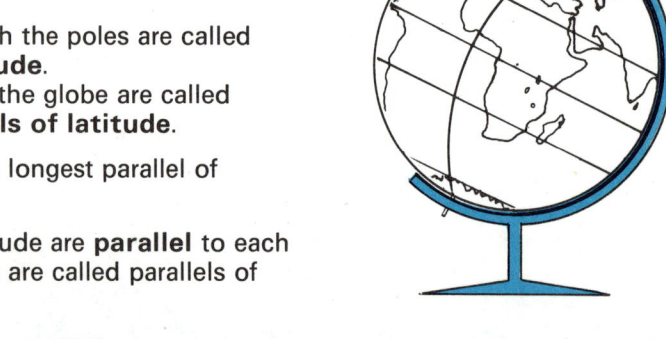

The lines through the poles are called
**lines of longitude**.
The lines round the globe are called
**lines** or **parallels of latitude**.

1.  Which is the longest parallel of
    latitude?

    Lines of latitude are **parallel** to each
    other and so are called parallels of
    latitude.

●   Here are two parallel lines. They never
    intersect no matter how much of
    them you draw.

2.  This match-box has two parallel
    edges marked in blue.

    **a.**  Point to a third edge which is
    parallel to the two blue lines.

    **b.**  If the match-box is opened a little
    more, will the two blue lines still be
    parallel?

    **c.**  If the match-box is nearly closed
    will the two blue lines still be
    parallel?

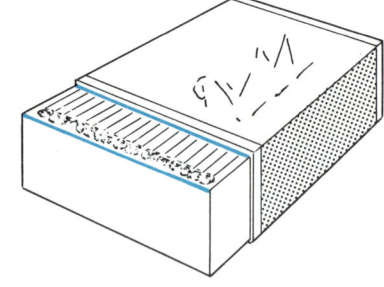

In this diagram, lines p and q are
parallel, and lines q and r intersect at
right angles.

3.  Check that lines p and r intersect at
    right angles.

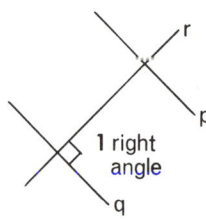

4.  Measure the distance:

    **a.** A to B   **b.** C to D   **c.** P to Q   **d.** R to S

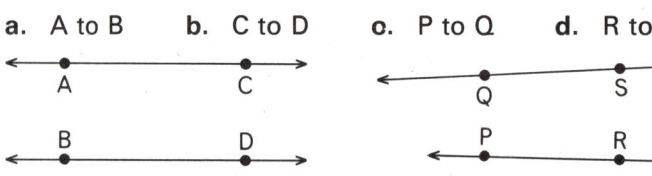

    **e.**  Which of the two pairs of lines are parallel?

To measure the distance between two parallel lines, draw a segment EF which is at right angles to the lines, and measure its length.

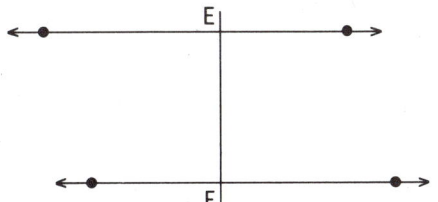

**5.** Check that the distance between these parallel lines is 2 centimetres.

**Let's Try**

**1. a.** Name each of these quadrilaterals:

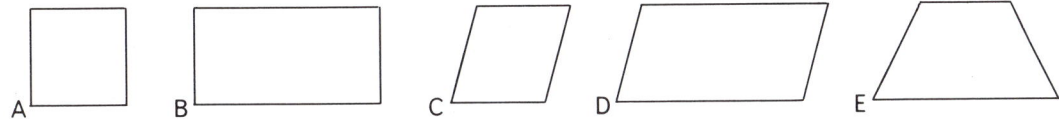

    **b.** Which of these quadrilaterals has only one pair of parallel sides?

    **c.** Which of these quadrilaterals has two pairs of parallel sides?

**2. a.** Measure the distance between these parallel lines.

    **b.** Trace the two lines and then draw two line segments to form a rectangle.

    **c.** Trace the two lines and then draw two line segments to form a parallelogram.

**3.** Use an atlas to find the parallel of latitude on which **Tewkesbury** lies.

**4. a.** Draw two parallel lines which are the width of your ruler apart.

    **b.** Draw two parallel lines which are 5 cm apart.

**5. a.** Cut two strips of paper 20 cm by 3 cm. Tie a reef knot: "Left over right and then right over left." Cut along the segments AB and CD.

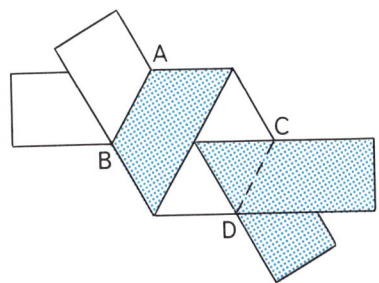

    **b.** You have formed a hexagon. How many pairs of parallel edges has a hexagon?

**6.** How many pairs of parallel edges can you find on a cube?

11

**Venn Diagrams**

Joe Brown had three sons and eight grandchildren; he drew his family tree like this.

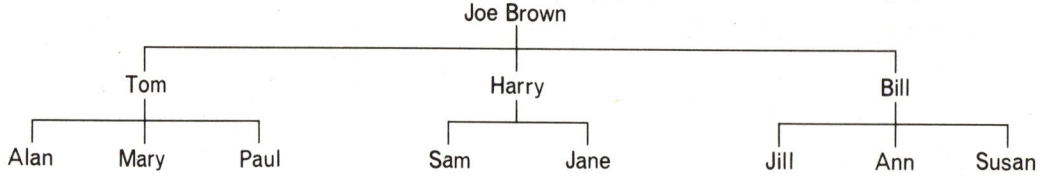

1. Complete these sets:

   a. The set of grandsons of Joe, S = {Alan, . . .

   b. The set of grand-daughters of Joe, D = {Mary, . . .

   c. The set of male descendants of Joe, M = {Tom, Paul, . . .

   Joe decided to show his sets in **regions**.
   M = {Male descendants}
   G = {Grandchildren}

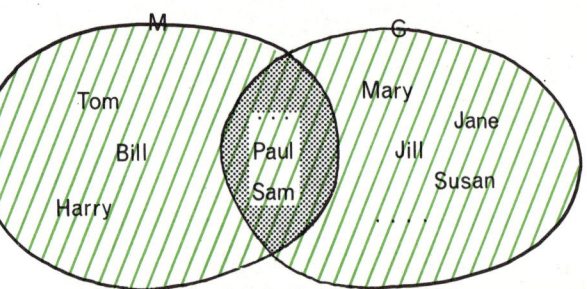

2. Write M ∩ G = {Sam, . . . and fill in the missing name in the grey region.

● The overlap of the region M and the region G contains elements which belong to both M and G, that is the **intersection** of M and G. The sign ∩ means "is the intersection of".

3. Write M ∪ G = {Tom, Paul, Mary, . . . } and fill in the missing name in the green region.

● The green region on the diagram contains elements which are members of M, or G. The elements may belong to both M and G. The green region represents the **union** of M and G. The sign ∪ means "is the union of".

4. From this diagram list the set:

   a. P = { . . . }   b. Q = { . . . }

   c. P ∩ Q = { . . . }   d. P ∪ Q = { . . . }

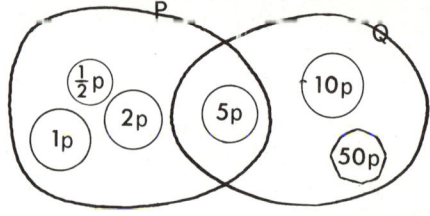

● Diagrams which represent sets are called **Venn diagrams**.

   P = {coins less than 10p}   Q = {silver coins}

5. Just by looking at the diagram can you see that:

   a. there is one silver coin less than 10p?

   b. there are six coins altogether?

This is part of the Venn diagram showing P and Q. The elements in this region are **members** of P but **not members** of Q.

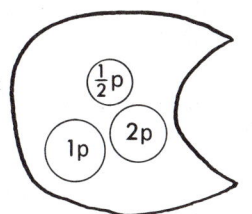

6. Check that elements in this region form coins less than 10p which are not silver.

7. Copy the Venn diagram and shade the region which forms {silver coins which are greater than 5p}.

Let's Try

1. From {5, 10, 15, 20, 25, 30, 35} list:

   a. P = {multiples of 10}
   b. Q = {multiples of 3} and show P and Q on the Venn diagram.
   c. P ∪ Q          d. P ∩ Q
   e. the set of multiples of 10 which are not multiples of 3.
   f. the set of multiples of 3 which are not multiples of 10.

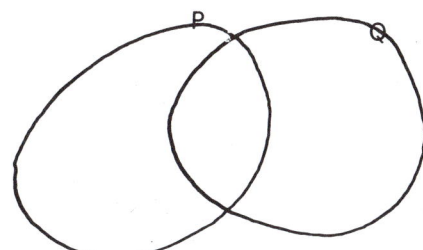

2. The Smith family has 6 children whose names are shown in set S. The Brown family has 5 children whose names are shown in set B.

   List the following set:

   a. S          b. B
   c. S ∩ B      d. S ∪ B
   e. Christian names which are not used by Mrs. Brown.
   f. Christian names which are not used by Mrs. Smith.

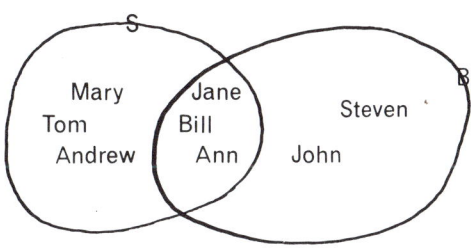

3. a. Enter on this diagram:
      the set of vowels (V), and the set of letters of our alphabet (A).
   b. Write A or V as the missing set in:
      A ∪ V = . . .     A ∩ V = . . .
   c. Describe the set of letters which belong to V but not A.

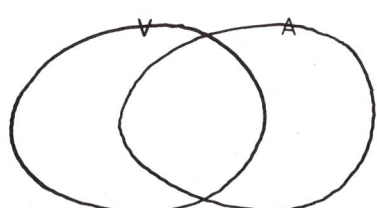

13

**Factors and Divisibility**

**1.** Find which one of the following numbers is not **divisible** by 6:

a. 36    b. 54    c. 78    d. 103

6 is a **factor** of 54 because there is no remainder when 54 is divided by 6, so 54 is **divisible** by 6.
103 has no factors other than 1 and itself and is called **prime**

**2.** Test each of the following rules for divisibility.

- **a.** A number ending in 0 or 5 is divisible by 5.
- **b.** A number ending in 0 is divisible by 10.
- **c.** 2 is a factor of every even number, and 2 is not a factor of an odd number.

**3.** A = {factors of 20} = {1, 2, 4, 5, 10, 20}
B = {factors of 12} = {1, 2, 3, 4, 6, 12}

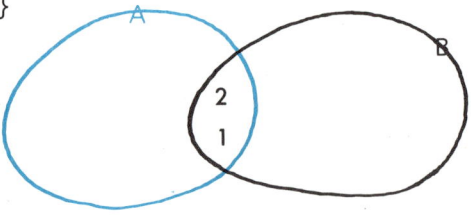

a. Copy this diagram and enter on it all the elements from A and B.

b. List A ∩ B = {1, 2, . . . }

c. Which is the largest number from A ∩ B?

The intersection (∩) of A and B is a list of factors which are factors of A and also factors of B.
These factors 1, 2, and 4 are called **common factors**.
4 is the **highest common factor** of 12 and 20.

**4. a.** Complete these sets and the diagram.
P = {factors of 15}
Q = {factors of 30}

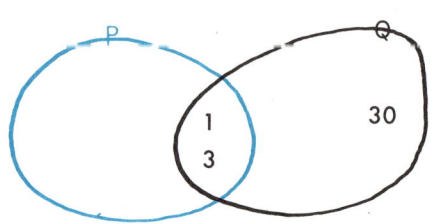

b. P ∩ Q = {1, . . . }

c. Is 15 the **highest common factor** of 15 and 30?

Sometimes the highest common factor of two numbers is the lower of the two numbers.

You have learned how to use factors to help in multiplication.
For example: 24×18 =  24×(6×3)
$$= (24×6)×3 = 144×3$$
$$= 432$$

Factors can help you divide.

5. Check and complete this working:

$$432 \div 18 = 432 \div (2 \times 9) = (432 \div 2) \div 9$$
$$= 216 \div 9 = \ldots$$

To divide by 18 we can divide first by 2 and then by 9 since $2 \times 9 = 18$.

## Let's Try

1. Write the set of factors of each of the following numbers:

   a. 14    b. 28    c. 50    d. 100

2. Write the set of prime numbers by which each of the following is divisible:

   a. 16    b. 81    c. 60    d. 100    e. 43

   f. Which element of {16, 81, 60, 100, 43} is prime?

3. For each of the following pairs of numbers:
   list the set of factors of the numbers,
   list the intersection of the two sets, and
   write the highest common factor of the two numbers.

   a. 15, 50    b. 27, 36    c. 100, 75    d. 43, 34
   e. 31, 103    f. 37, 111    g. 100, 200    h. 129, 86

4. Complete these multiplications and divisions by making use of factors:

   a. $47 \times 14$    b. $121 \times 16$    c. $478 \times 15$
   d. $492 \div 12$    e. $574 \div 14$    f. $1248 \div 16$

5. **Twin primes** are prime numbers which differ by 2.
   For example, 5 and 7 are "twin primes".
   Find a pair of "twin primes" between 95 and 108.

6. Andrew is trying to find the highest common factor of 89 and 97. Barry said,
   "That's easy; the two numbers are prime so the highest common factor is 1."
   Is Barry's rule correct?

7. Andrew thinks that he has found a rule for testing divisibility by 3.
   "Check whether the sum of the digits is divisible by 3."
   For example $\boxed{135}$ $1+3+5 = 9$; 9 is divisible by 3, and so 135 is divisible by 3.

   Try this rule with some numbers other than 135.

On a long journey, you may have played games such as "John counts the red cars and Jane counts the blue cars."
If you need to keep a count of cars, lorries, vans, . . . then use the **tally method**. Count in groups of five so that the final totalling is easier.

Tally

| I | II | III | IIII | ЦΉ | ЦΉ I | ЦΉ II | ЦΉ ЦΉ | ЦΉ ЦΉ I |
|---|----|-----|------|-----|-------|--------|--------|----------|
| 1 | 2  | 3   | 4    | 5   | 6     | 7      | 10     | 11       |

You make a mark each time a vehicle passes. For every fifth one you draw a line through the previous four marks. ЦΉ ЦΉ ЦΉ ЦΉ I I I = 4 fives+3 = 23.

**1.** Check the following tallies. Each tally is an element of {33, 32, 37}.

    **a.** ЦΉ ЦΉ ЦΉ ЦΉ      **b.** ЦΉ ЦΉ ЦΉ ЦΉ      **c.** ЦΉ ЦΉ ЦΉ
        ЦΉ ЦΉ II           ЦΉ ЦΉ ЦΉ II        ЦΉ ЦΉ ЦΉ III

**2.** Here is a list of shoe sizes for a group of children:
    5, 3, 4, 5, 6, 4, 4, 6, 5, 5, 3, 4, 5, 5, 5, 4, 3, 3, 5, 4

The tally has been started by making a mark for 5, then 3, then 4, then 5.
The **total** number of size 5 shoes has been entered for you in the box opposite 5.
Copy and complete the **tally table** for shoe size and then draw a column graph from the information.

| Shoe size | Number of children | |
|-----------|--------------------|--|
| 3 | I | |
| 4 | I | |
| 5 | II | 8 |
| 6 | | |
| | | 20 |

The total number in each row is called the **frequency**.
For example, the frequency of size five shoes is 8.

Ask your parents if they have ever taken part in a survey. It might have been a survey to find out how many people in your town watched TV between 19.00 hours and 20.00 hours.
After speaking to 100 people, an **interviewer** found that:

19.00 → 20.00 hours

| Viewing | ЦΉ ЦΉ ЦΉ ЦΉ ЦΉ ЦΉ ЦΉ ЦΉ ЦΉ ЦΉ ЦΉ ЦΉ ЦΉ III |
|---------|-----------------------------------------------|
| Not viewing | ЦΉ ЦΉ ЦΉ ЦΉ ЦΉ ЦΉ II |

**3.** How many people were:   **a.** viewing    **b.** not viewing
    Do your answers to **a** and **b** add up to 100?

Of the people interviewed: the decimal fraction viewing was 0·68, and the decimal fraction not viewing was 0·32.

4. There are 10000 people in your town. Complete this working:

   **a.** Total number viewing
   $= 0·68 \times 10000$
   $= \ldots$

   **b.** Total number not viewing
   $= 0·32 \times 10000$
   $= \ldots$

It is **not** true to say that **exactly** 3200 were not viewing.
The figures are only approximate but they give us a guide and enable us to say that:
"**About** 7000 were viewing and **about** 3000 were not."

Let's Try

1. This list of numbers shows the ages in years of a group of children:
   11, 9, 10, 12, 11, 11, 9, 12, 10, 12, 9, 12, 11, 9, 11, 10, 12, 9, 10, 12, 10, 11, 9, 12, 10, 12, 10, 11, 11, 12, 11, 9, 12, 11, 11, 10, 9, 10, 11, 12, 11, 9, 10, 11, 12, 10, 11, 9, 10, 11, 10, 11, 11.

   **a.** Tally the number of children of age 9, 10, 11, and 12.
   How many children of each age are there?
   How many children are there altogether?

   **b.** Make a tally and draw a graph of the ages in months of the children in your class.

2. Weigh three lots of 50 grammes of dried peas, and count the number of peas in each weighing.

   | Weighing | 1 | 2 | 3 |
   |---|---|---|---|
   | Number of peas | | | |

   **a.** About how many peas would you say there are in 50 grammes?

   **b.** Work out the approximate number of peas in 1 kilogramme.

3. Five children in a school of 500 were asked, "Do you read *Toddlers Annual*?"
   2 answered, "No", and 3 answered, "Yes".
   **From this information** is it true to say that:

   **a.** $\frac{3}{5}$ of the children read *Toddlers Annual*?

   **b.** $\frac{2}{5}$ of the children do not read *Toddlers Annual*?

   **c.** Is it likely that our interviewing has produced reasonably correct information for the whole school? Think about the possible ages of the children. Do you think that more children should have been interviewed?

**Addition and Subtraction of Fractions**

1. Which of these fractions are equivalent to $\frac{1}{2}$?

$\frac{2}{1}$     $\frac{4}{2}$     $\frac{2}{3}$     $\frac{2}{4}$     $\frac{9}{19}$     $\frac{5}{10}$

● Fractions which represent the same number are called **equivalent** fractions.

You will remember adding and subtracting fractions by changing denominators. Complete the next example in order to refresh your memory.

2. This shunting engine travelled $\frac{3}{10}$ km then reversed $\frac{1}{5}$ km.

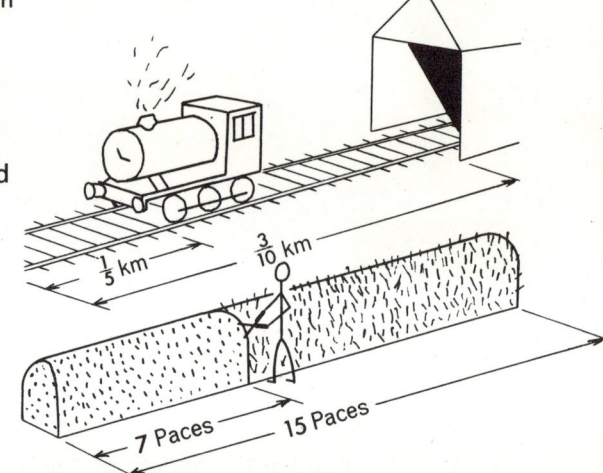

   a. Altogether the engine travelled
$$\frac{3}{10}+\frac{1}{5} = \frac{?}{10}+\frac{?}{10} = \frac{?}{10} = \frac{?}{2}\,km$$

   b. The distance the engine finished from its shed is
$$\frac{3}{10}-\frac{1}{5} = \frac{?}{10}-\frac{?}{10} = \frac{?}{10}\,km$$

3. John offered to cut the hedge. He knows that he has cut $\frac{7}{15}$; he completed this working to find the fraction left:
$$1-\frac{7}{15} = \frac{15}{15}-\frac{?}{?} = \frac{8}{?}$$

● Whole numbers can be written as fractions $1 = \dfrac{n}{n}$

Here are two mixed numbers to be added together: $\boxed{3\frac{1}{2}+1\frac{3}{5}}$

4. a. Is it true that $3\frac{1}{2}+1\frac{3}{5} = 3+\frac{1}{2}+1+\frac{3}{5} = 3+1+\frac{1}{2}+\frac{3}{5} = 4+\frac{1}{2}+\frac{3}{5}$?

● When adding mixed numbers, add the whole numbers first and then add the fractions.

   b. Complete the addition:
$$3\frac{1}{2}+1\frac{3}{5} = 4+\frac{1}{2}+\frac{3}{5} = 4+\frac{?}{10}+\frac{?}{10} = 4+\frac{?}{10} = 4+1\frac{1}{10} = \ldots$$

Now let us learn how to subtract $1\frac{2}{5}$ from $2\frac{1}{2}$.   $\boxed{2\frac{1}{2}-1\frac{2}{5}}$

5. a. Is it true that $8-(4+3) = 8-4-3$?

   b. Is it true that $2\frac{1}{2}-1\frac{2}{5} = 2+\frac{1}{2}-(1+\frac{2}{5}) = 2+\frac{1}{2}-1-\frac{2}{5} = 1+\frac{1}{2}-\frac{2}{5}$?

● When subtracting a mixed number, subtract the whole number and then subtract the fraction.

**c.** Check this working: $2\frac{1}{2}-1\frac{2}{5} = 1+\frac{1}{2}-\frac{2}{5} = 1+\frac{5}{10}-\frac{4}{10} = 1\frac{1}{10}$

**6.** This example is not quite so easy. Check through the working.

$$3\frac{1}{2}-1\frac{3}{5} = 2+\frac{1}{2}-\frac{3}{5} = 2+\frac{5}{10}-\frac{6}{10} = ?$$

In order to complete this example change a whole number to tenths. Here is the completed working:

$$3\frac{1}{2}-1\frac{3}{5} = 2+\frac{5}{10}-\frac{6}{10} = 1+\frac{10}{10}+\frac{5}{10}-\frac{6}{10} = 1+\frac{15}{10}-\frac{6}{10} = 1\frac{9}{10}$$

Let's Try

**1.** Add the following:

   **a.** $1\frac{1}{2}+2\frac{3}{10}$    **b.** $2\frac{7}{10}+1\frac{1}{2}$    **c.** $1\frac{1}{4}+\frac{9}{10}$

   **d.** $1\frac{3}{5}+2\frac{1}{4}$    **e.** $1\frac{2}{5}+\frac{3}{4}$    **f.** $2\frac{1}{4}+1\frac{1}{2}+1\frac{3}{5}$

**2.** Complete these subtractions:

   **a.** $2\frac{1}{2}-1\frac{3}{10}$    **b.** $2\frac{3}{10}-\frac{1}{2}$    **c.** $3\frac{2}{5}-1\frac{7}{10}$

   **d.** $10\frac{1}{4}-5\frac{3}{5}$    **e.** $4\frac{3}{4}-3\frac{9}{10}$    **f.** $5\frac{1}{10}-2\frac{3}{4}$

**3.** This fishing rod is made up of 3 sections:
A = $\frac{3}{4}$ metre, B = $\frac{7}{10}$ metre, C = $\frac{2}{5}$ metre

Work out the total length in metres of:

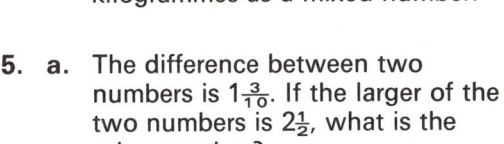

   **a.** Sections A and B

   **b.** Sections B and C

   **c.** Sections A and C

   **d.** Sections A, B, and C

**4.** Using this set of scales one box of strawberries weighs 500 g and the other 600 g.

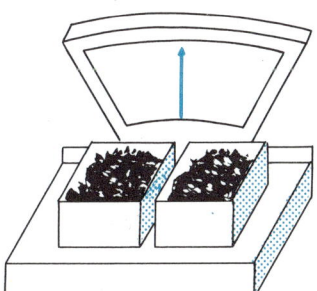

   **a.** Write the weights as fractions of a kilogramme.

   **b.** Add your two fractions from part **a** and write the total weight in kilogrammes as a mixed number.

**5.**  **a.** The difference between two numbers is $1\frac{3}{10}$. If the larger of the two numbers is $2\frac{1}{2}$, what is the other number?

   **b.** The sum of two numbers is $3\frac{7}{10}$. If one of the numbers is $1\frac{1}{5}$ what is the other number?

A **main road** passes through Stafford, Stone, Newcastle-under-Lyme, and Congleton.

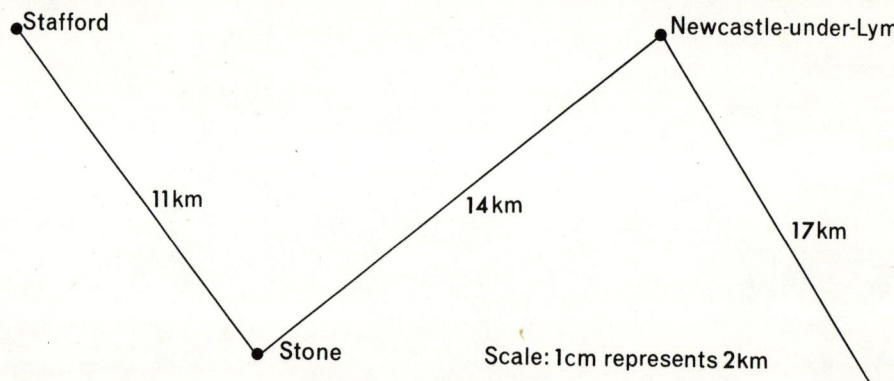

Scale: 1cm represents 2km

The distance in kilometres between towns is shown on the route map.

1.  **a.** Measure the length of the line segment between Stafford and Stone.
    Do 5·5 cm represent 11 km? Check with the scale.

    **b.** Use the map to check the distance between:
    Stone and Newcastle-under-Lyme, and Newcastle-under-Lyme and Congleton.

    Here is a table which
    needs completing.

| Stafford | | | |
|---|---|---|---|
| 11 | Stone | | |
| 25 | 14 | Newcastle-under-Lyme | |
| | | 17 | Congleton |

2.  **a.** Look at the entry shaded green. Does 11+14 = 25?
    Is it 25 km from Stafford to Newcastle-under-Lyme on the main road?

    **b.** The box shaded grey shows the distance from Stone to Congleton.
    Copy the table and fill in the missing number.
    14+17 = . . .

    **c.** Write the missing number in the box which has been left white.
    This is the distance between . . . and . . . .

Since the main road passes through all four towns, the missing distance can easily
be found.

Here is a new map showing the distance between towns along main roads.

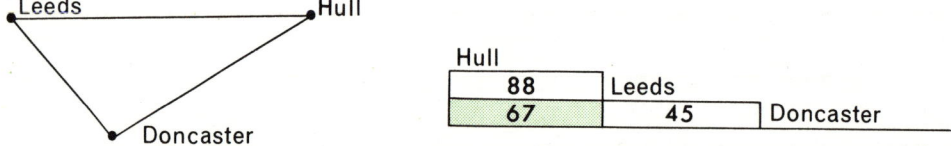

| Hull | | | |
|---|---|---|---|
| 88 | Leeds | | |
| 67 | 45 | Doncaster | |

3.  How far is it from Hull to Doncaster if you call at Leeds on the way?

| Exeter | Manchester | Leeds | Stoke-on-Trent | Northampton | London | Nottingham | York |
|---|---|---|---|---|---|---|---|
| 376 | | | | | | | |
| 430 | 64 | | | | | | |
| 318 | 59 | 117 | | | | | |
| 288 | 190 | 198 | 136 | | | | |
| 272 | 294 | 304 | 235 | 104 | | | |
| 440 | 112 | 107 | 80 | 91 | 192 | | |
| 467 | 102 | 38 | 160 | 216 | 314 | 122 | |

Kilometres

**1.** From the table find the distance from:

  **a.** Exeter to London

  **b.** Manchester to London

  **c.** Leeds to York

  **d.** York to London

**2.** Which town is:

  **a.** 294 km from London

  **b.** 102 km from Manchester

  **c.** 190 km from Manchester

  **d.** 160 km from Stoke-on-Trent

**3.** **a.** Make up a "distance" table from this route map:

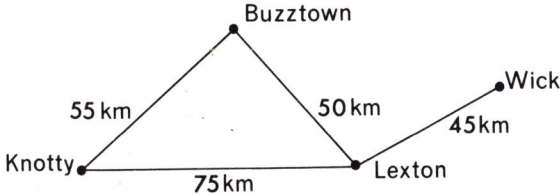

  **b.** At 60 km per hour, how long will it take to travel from:
  Knotty to Wick; and from Knotty to Wick via Buzztown.

**4.** Look at a "distance" table in an *Automobile Association Handbook*. Use it to find the distance between London and Birmingham.

  **a.** Plan a car journey from your home. Arrange to pass through four cities. Draw a route map and make a "distance" table.

  **b.** From a map or atlas find out how far your destination is from home by air.

**Binary Numbers**

This box contains 25 peaches.

1. **a.** Write the missing number,
   $25 = 5 \times \ldots$

   **b.** Is the box square?

   25 is a square number. It is the
   **square** of 5.
   $25 = 5 \times 5 = 5^2$, or **five** to the
   **power two**

2. Complete this pattern:

$$5 \times 5 = 5^2 = 25$$
$$5 \times 5 \times 5 = \ldots = 125$$
$$5 \times 5 \times 5 \times 5 = 5^4 = \ldots$$

Using **powers** can often save a lot of writing. For example:
in 1 kilometre there are 100000 centimetres
or $10^5$ centimetres
$10^5 = 100000$

3. Copy these tables and fill in the missing numbers:

| $2^6$ | $2^5$ | $2^4$ | $2^3$ | $2^2$ | $2^1$ |
|---|---|---|---|---|---|
|  |  |  |  |  | 2 |

| $10^4$ | $10^3$ | $10^2$ | $10^1$ |
|---|---|---|---|
|  |  |  |  |

You might think that the cleverest boy
in your class is quick at arithmetic.
Could he find $235479 \times 7452163$
in 1 second? This computer could.
You will have heard about computers.
They are machines which can do
calculations for us.

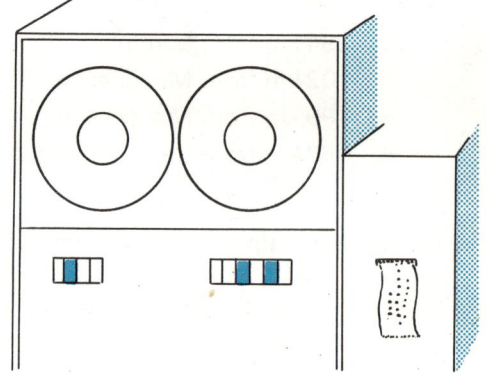

Here are the column headings for
computer arithmetic which works in
**base two**:

| Sixteens | Eights | Fours | Twos | Units |
|---|---|---|---|---|
| $2^4$ | $2^3$ | $2^2$ | $2^1$ | 1 |

● The column values are **powers** of 2. **Base two** numbers are called **binary**
numbers, and we write $_{(2)}$ at the foot of each number.

4. Fill in the missing parts of this table.

| Binary Number | | | | | | Base Ten |
|---|---|---|---|---|---|---|
| Sixteens | Eights | Fours | Twos | Units | | |
|  |  |  |  | $1_{(2)}$ |  | 1 |
|  |  |  | 1 | $0_{(2)}$ | $= (1 \times 2) =$ | 2 |
|  |  |  | 1 | $1_{(2)}$ | $= (1 \times 2) + 1 =$ | 3 |
|  |  | 1 | 0 | $0_{(2)}$ |  |  |
|  |  | 1 | 0 | $1_{(2)}$ |  |  |
|  |  | 1 | 1 | $0_{(2)}$ |  |  |
|  |  | 1 | 1 | $1_{(2)}$ |  |  |

The next binary number is $1000_{(2)} = 8_{(10)}$

**5.** Try counting with binary numbers.
$1001_{(2)}$ is read as "one nought nought one **base two**".
Our base ten numbers can be changed to **base two** by splitting into
. . . 32s, 16s, 8s, 4s, 2s, and units.

**6.** Check this working:
$53_{(10)} = 32+21 = 32+16+5 = 32+16+0+4+0+1$

so $53_{(10)} = 110101_{(2)}$

## Let's Try

**1.** Change these binary numbers to base ten:

   **a.** 1101   **b.** 11010   **c.** 11011   **d.** 1010111

**2.** Here is a piece of "punched tape" which is used by a computer. **0** is a punched hole and 0 is not punched. Number **a** is $11110_{(2)}$

Write the binary numbers shown on the tape and change them to base ten.

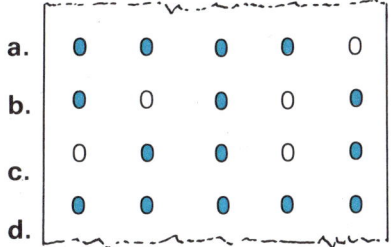

**a.**
**b.**
**c.**
**d.**

**3.** Answer each of these questions by counting in binary: "one", "one nought", "one one", . . .
Check your answers by changing them to base ten.

   **a.** How many toes have you?

   **b.** Count the number of girls in your class.

   **c.** How many boys are there in your class?

   **d.** What is the total number of children in your class?

   **e.** How many days are there in a week?

**4.** Change these base ten numbers to base two.

   **a.** 32   **b.** 48   **c.** 96   **d.** 112

   **e.** $2^4$   **f.** $2^5$   **g.** $2^3+2^4$   **h.** 100

Here is an interesting **series** of numbers.

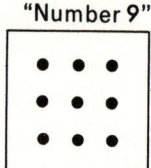
"Number 9"

- 1, 4, 9, 16, 25, . . .
- This is called the **series of square numbers**.

Each number in the series can be represented by a square array.

**1.** **a.** Complete this table:

| △ | 1 | 2 | 3 | | | 6 |
|---|---|---|---|---|---|---|
| △ × △ | | 4 | | 16 | 25 | |

**b.** Copy these parallel number lines and then join "a number" to its "square". For example, 2 to 4:

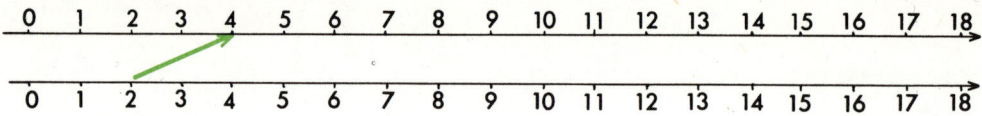

You will need to extend the number lines to 36.

**c.** Did you notice how quickly your arrow approached the direction → ? Will it ever become horizontal → ?

Your two number lines have been used to graph the set of ordered pairs, {(1, 1), (2, 4), (3, 9), (4, 16), . . .}
Two number lines can be used to graph a series. For example:
0, 1, 4, 5, 8, 9, 12, 13, 16, . . .

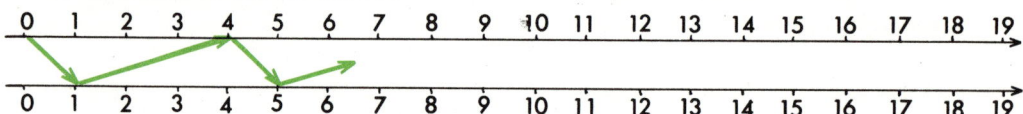

**2.** Copy and complete the graph up to 16.

Have you noticed how shopkeepers display their wares?

**3.** How many oranges are there in this pile?
What is the shape of the display?

- A number which can be represented by a triangular array is called **triangular**.

**4.** **a.** Here is a series of **triangular numbers**. Write the "differences".

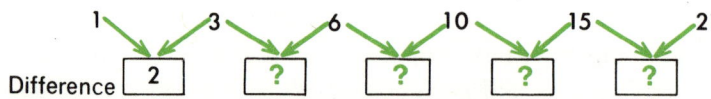

1    3    6    10    15    21

Difference | 2 | ? | ? | ? | ? |

**b.** Write the next three triangular numbers.

**c.** Complete this arrow graph for triangular numbers up to 36.

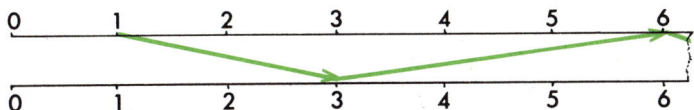

**1.** Draw two number lines and arrow graphs for each of the following sets of ordered pairs.

   **a.** {(1, 5), (2, 6), (3, 7), (4, 8), (5, 9), (6, 10)}

   **b.** {(0, 7), (1, 6), (2, 5), (3, 4), (4, 3), (5, 2), (6, 1), (7, 0)}

   **c.** {(5, 0), (6, 1), (7, 2), (8, 3), (9, 4), (10, 5)}

   **d.** {(1, 2), (2, 4), (3, 6), (4, 8), (5, 10), (6, 12)}

**2.** Draw two parallel number lines up to 30. Write the next two numbers in each of the following series and then draw arrow graphs.

   **a.** 1, 5, 7, 11, 13, 17, 19, 23, . . .    **b.** 5, 10, 15, 20, . . .

   **c.** 2, 6, 10, 14, 18, 22, . . .      **d.** 30, 0, 25, 5, 20, . . .

**3.** 1, 3, 6, 10, 15, 21, 28, 36, 45, 55

   **a.** Check that these are triangular numbers.

   **b.** Try adding two consecutive numbers in the series. For example, 3+6. What "shape" is your new number?

   **c.** Look at this working.

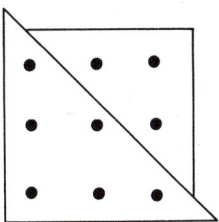

● Half the product of any two consecutive numbers gives a triangular number.

$$\frac{3 \times 4}{2} = 6$$

Check this rule.

**4.** Can you see the pattern in this series? 2, 4, 7, 11, 16, . . .
Draw an arrow graph of this series on the number lines you used for the triangular numbers. Use a different colour.

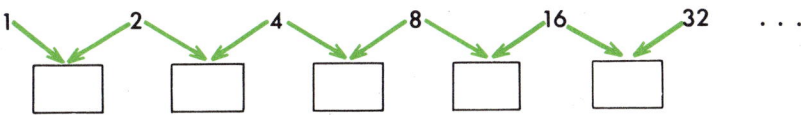

Fill in the "differences" and then draw an arrow graph of the series.

Susan is a lucky girl. She has been given a 50p piece by her uncle.
She decides to spend only 2p each day and makes up a table like this:

| Days of spending | 0 | 1 | 2 | . . . |
|---|---|---|---|---|
| Amount left | 50p | 48p | 46p | . . . |

**1.** **a.** Complete the table for Susan.

**b.** How much will Susan have
after: 5 days, 10 days,
25 days?

It would have been
better if Susan had
drawn a graph like this.

**2.** **a.** Draw this graph for
yourself using a
larger scale.

**b.** Did you mark all 25
points from Susan's
table before you
realized that the
graph is a straight line?

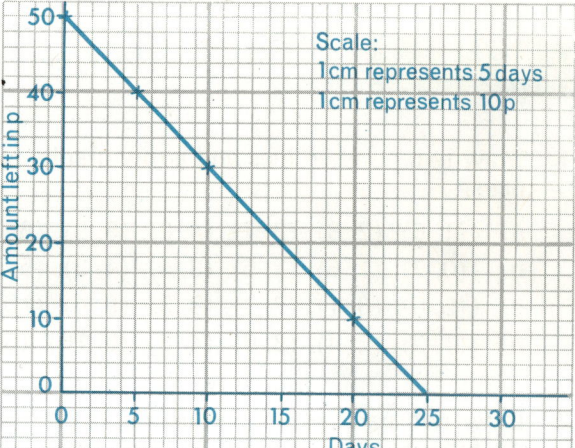

Scale:
1cm represents 5 days
1cm represents 10p

● When drawing a graph which is a straight line, you need mark only 2 points.
Usually it is safer to mark 3 points in case one is wrong.

Susan's money is being spent at a
**constant rate**.

Here is a picture of a **gasometer**. As
it fills with gas, it rises from the ground.
Gas is being pumped in steadily so
that the container rises 0·5 metre
per day.
This **graph showing the height
of the gasometer** has been started
for you.

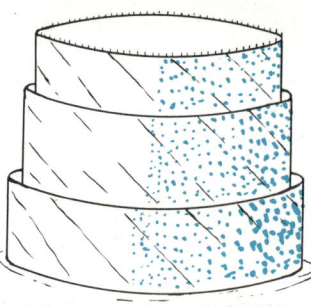

**3.** **a.** Draw the graph for yourself.
Use this table.

| Height after | 0 | 5 | 10 | metres |
|---|---|---|---|---|
| | 0 | 10 | 20 | days |

**4.** Check from this graph that:

**a.** The height after 3 days is 1·5
metre.

**b.** The gasometer is 2 metres high
after 4 days.

**5.** **a.** When would this graph end?
Could it go on for ever?

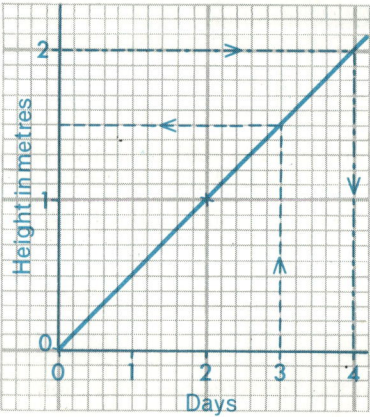

1. Here is a **graph for converting £ to francs, and francs to £**. The graph shows that £1 = 12 francs.

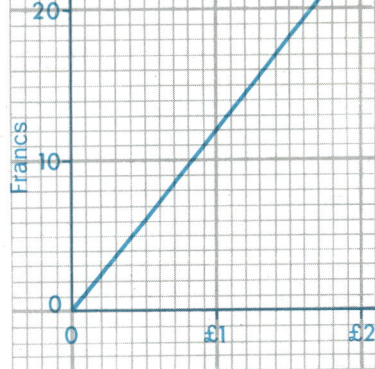

   a. What scale has been used on the £ axis?

   b. What scale has been used on the francs axis?

   c. How many francs can be exchanged for:
   50p; 150p; £1·25?

   d. How many p can be exchanged for:
   9 francs; 21 francs; 10 francs?

2. John's father says that he will give him 1p on the 1st July, 2p on the 2nd July, 3p on the 3rd July, . . . and so on until the 21st July.

   a. How much did John receive on the 21st July?

   b. Draw a graph showing the amount received on each date.

   c. Complete this table:

   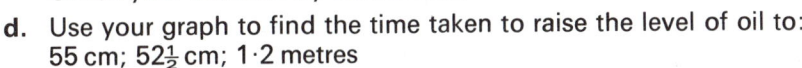

| Date | 1 | 2 | 3 | 4 | 5 |
|------|---|---|---|---|---|
| Amount | | | | | |

   **Total** = . . . p

   d. How much did John receive altogether?

3. The Oil Company tanker is filling a storage tank which is used to feed a central heating system.
   The depth of oil in the tank is rising steadily at a rate of 10 cm per minute.

   a. If the tank is 1·5 metres high, how long does it take to fill the tank?

   b. Draw a graph to show the way in which the depth of the oil changes.

   c. Use your graph to find the depth after:
   7 minutes; $7\frac{1}{2}$ minutes; $8\frac{1}{4}$ minutes
   Check your answers by calculation.

   d. Use your graph to find the time taken to raise the level of oil to:
   55 cm; $52\frac{1}{2}$ cm; 1·2 metres

   Check your answers by calculation.

**Some Special Factors**

John has been juggling with the set of
natural numbers, {1, 2, 3, 4, . . . }.
He is carefully catching some numbers
such as {3, 6, 9, 12, 15, 18, 21, . . . }
and letting the rest fall.

1.  a.  Check that John's rule is:
        "Catch if the number is divisible
        by 3". For example, $18 \div 3 = 6$
        **remainder 0**.

    b.  I think that his rule should be:
        "Catch if the sum of the digits is
        divisible by 3".
        For example, 18 is $(1+8) = 9$.
        Does my rule work for 18?

    c.  Should John catch 48 according
        to **my** rule?

● If the sum of the digits is divisible by three, then the original number is also divisible
  by three.

Here is a factor tree for 12.

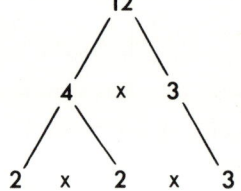

2.  a.  Is $12 = 2 \times 2 \times 3$?

    b.  Are there any factors of 2 other
        than 1 and 2?

    2 and 3 are **prime numbers**.

● A prime number has no factors other than 1 and itself.

2 is a special **factor** of 12 and is called a **prime factor**.

3.  Complete this sentence:

    a.  "4 is a factor of 12 but is **not** a . . . **factor**".

    b.  Write the missing prime factor of 12 in {1, 2, . . . }.

● Prime factors of a number appear in the last line of its factor tree.

Check this on the factor tree for 12.

4.  Write a factor tree for 60.
    Your last line should equal $2 \times 2 \times 3 \times 5$.
    Remember that $3 \times 2 \times 2 \times 5 = 2 \times 5 \times 2 \times 3 = 2 \times 2 \times 3 \times 5$.

    60 has been **completely factorized** when written as the product of prime factors.
    $60 = 2 \times 2 \times 3 \times 5$.

**5.** $\boxed{2 \times 2 \times 3 \times 5} = 60$   Take two numbers from the box and find the product of them.

For example, $2 \times 2 = 4$.
Is your product a factor of 60? Try two more numbers from the box.

- From the set of factors {2, 2, 3, 5} the product of any two factors, or any three factors, is also a factor of 60.

**6.** Check that the product of 2, 2, 3, and 5 is a factor of 60.

**Let's Try**

**1.** Cook has prepared 246 cakes and intends wrapping them in packs of 3.

   **a.** Will there be any cakes left over?

   **b.** Would 1020 cakes pack exactly into 3s?

**2. a.** Write the first ten numbers which are divisible by 5.

   **b.** Make up a rule for finding numbers which have 5 as a factor.

**3.** Complete the following rules and then test them.

   **a.** An . . . number always has 2 as a factor.

   **b.** A number with 0 in the units column will have 2, 5, and . . . as factors.

**4.** Here is a "pair of primes" whose difference is 2: (11, 13).
List all the "pairs of primes" less than 100 with a difference of 2.

**5. a.** Completely factorize 36 by drawing a factor tree.

   **b.** Use the last line to help you list the set of prime factors of 36.

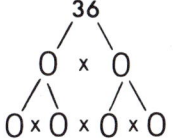

**6. a.** Find the sum of all the factors of 28 including 1 but not 28.
$1 + . . . + . . . + . . . + . . . = . . .$
Numbers with this property are called **perfect numbers**.

   **b.** Check that 6 and 496 are **perfect numbers**.

**7.** List the set of prime factors of:
   **a.** 20   **b.** 100   **c.** 250   **d.** 194

## How to Convert

1. Cut the following lengths from a piece of string or cotton:

   **a.** 50 cm  **b.** $\frac{1}{2}$ metre

   **c.** $\frac{5}{10}$ metre  **d.** 0·5 metre

   Are your pieces of string all the same length? They should be since
   $\frac{1}{2}$ m = $\frac{5}{10}$ m = 0·5 m = 50 cm

   **Remember that $\frac{1}{4}$ is 1÷4 or $4\overline{)1}$**

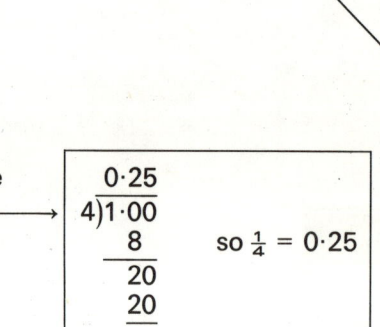

2. Complete this working:
   $$\frac{1}{4} = \frac{?}{100} = 0·25$$

   In order to change $\frac{1}{4}$ to a decimal write $4\overline{)1·00}$ and divide like this:———————➤

   $$\begin{array}{r} 0·25 \\ 4\overline{)1·00} \\ 8 \\ \hline 20 \\ 20 \\ \hline \end{array} \qquad \text{so } \tfrac{1}{4} = 0·25$$

3. Use this method of dividing the numerator by the denominator to complete the following:

   **a.** $\frac{3}{5}$ = ...  **b.** $\frac{3}{4}$ = ...  **c.** $\frac{3}{20}$ = 0·?5

   $$5\overline{)3·0} \qquad\qquad 4\overline{)3} \qquad\qquad \begin{array}{r}0· \\ 20\overline{)3·00}\end{array}$$

   In order to change $\frac{1}{8}$ to a decimal, three decimal places will be needed.

   $\frac{1}{8}$ = 0·125

   $$\begin{array}{r} 0·125 \\ 8\overline{)1·000} \\ 8 \\ \hline 20 \\ 16 \\ \hline 40 \\ 40 \\ \hline \end{array}$$

4. Complete these two examples:

   **a.** $\frac{3}{8}$ = ...  **b.** $\frac{3}{500}$ = ...

   $$\begin{array}{r} 0·3?? \\ 8\overline{)3·000} \\ 2\,4 \\ \hline 60 \end{array} \qquad\qquad \begin{array}{r} 0·00? \\ 500\overline{)3·000} \end{array}$$

   Try changing $\frac{1}{3}$ to a decimal.

   $\frac{1}{3}$ = 0·33333

   $$\begin{array}{r} 0·33333 \\ 3\overline{)1·00000} \end{array}$$

**5.** Did you find that the division of 1 by 3 never ended?

0·33333                                          0·122222

are recurring decimals and are written as

0·3̇                                              0·12̇

**1.** Change these fractions to decimals:

a. $\frac{13}{100}$   b. $\frac{7}{20}$   c. $\frac{19}{20}$   d. $\frac{4}{25}$   e. $\frac{12}{25}$

f. $\frac{9}{50}$   g. $\frac{1}{40}$   h. $\frac{3}{40}$   i. $\frac{1}{8}$   j. $\frac{7}{8}$

**2.** Susan wishes to buy $\frac{3}{4}$ metre of curtain material. The shop assistant says,
"This piece will be suitable because it is 0·7 metres long."

a. Change $\frac{3}{4}$ to a decimal.

b. Write the difference between $\frac{3}{4}$ metre and 0·7 metre in centimetres, and as a decimal of a metre.

c. Is the piece of curtain material long enough for Susan?

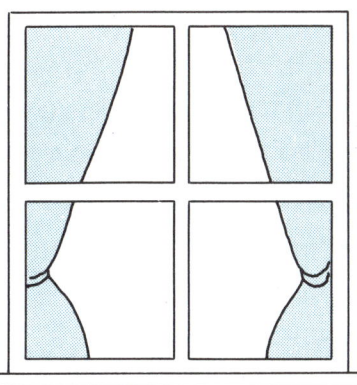

**3.** Change these fractions to decimals which recur:

a. $\frac{1}{9}$   b. $\frac{2}{9}$   c. $\frac{3}{9}$   d. $\frac{4}{9}$

e. $\frac{5}{9}$   f. $\frac{1}{6}$   g. $\frac{5}{6}$   h. $\frac{23}{90}$

**4.** Complete the following sentences using
> "is greater than", or < "is less than", or = "is equal to".

a. $\frac{3}{40}$   0·175   b. $\frac{1}{3}$   0·3   c. $\frac{1}{3}$   0·3̇

**5.** Try to change $\frac{1}{7}$ to a decimal. It is quite a puzzle.

This boy scout can walk 8 km in 1 hour.
His speed of walking is 8 km per hour.

1.  **a.** How far can he walk in 30
       minutes?

   **b.** How long does it take him to
       walk 12 km?

   To find how far he walks in a given
   time work like this:

   In 60 min he walks 8 km

   so in 1 min he walks $\frac{8}{60}$ km

   and in 12 min he walks $\frac{8}{\underset{10}{\cancel{60}}} \times \overset{2}{\cancel{12}}$ km

   = $\frac{16}{10}$ or 1·6 km

2.  Use this method to check your answer to question **1a**.

   To find out how long he takes to walk a given distance work like this:

   He walks 8 km in 60 min

   so      1 km takes $\frac{60}{8}$ min

   and      7 km take $\frac{\overset{15}{\cancel{60}}}{\underset{2}{8}} \times 7$ min

   = $\frac{105}{2}$ = $52\frac{1}{2}$ min

3.  Use this method to check your answer to question **1b**.

   Here is a **graph showing a walking speed of 6 km per hour**:

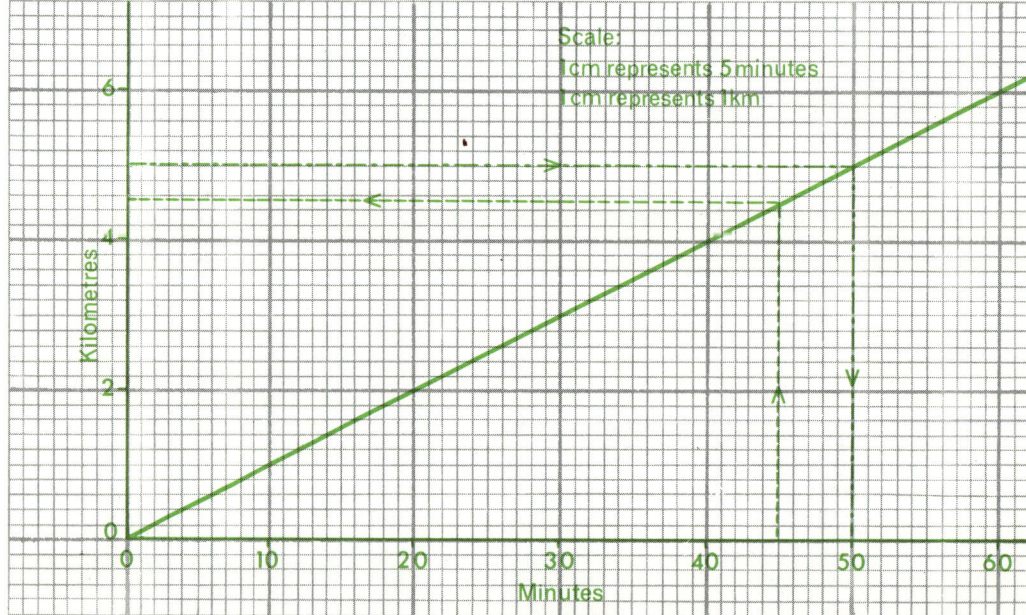

On the graph notice that:

**a.** the scale for each axis is shown, and

**b.** a title is given.

In order to find the distance travelled in 45 minutes we can read from the graph ↖↑. This distance is $4\frac{1}{2}$ km.

**4.** Check this answer by calculation.

In order to find the time taken to travel 5 km read ↘↓ from the graph. The time taken is 50 minutes.

**5.** Check this answer by calculation.

Let's Try

**1.** A car is travelling at 80 km per hour.
Use the method shown on the opposite page to work out the time taken to travel:

**a.** 24 km   **b.** 26 km   **c.** 60 km   **d.** 90 km

Use the method shown on the opposite page to work out the distance travelled in:

**e.** 45 minutes   **f.** 24 minutes   **g.** 35 minutes   **h.** 50 minutes.

**2.** Use the graph on the opposite page to write the distance travelled in:
**a.** 20 minutes   **b.** 35 minutes
**c.** 51 minutes

**3.** This **graph showing the speed of a cyclist** has been started for you.

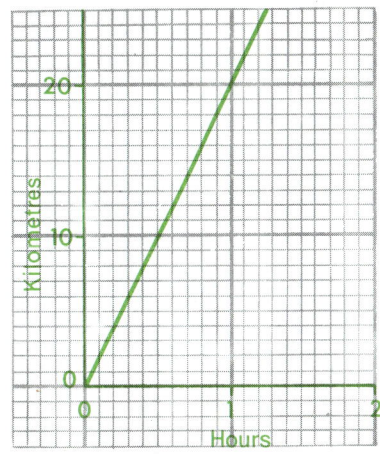

**a.** What scale has been used on the **time** axis?

**b.** What scale has been used on the **distance** axis?

**c.** At what speed is the cyclist travelling?

**d.** Draw the complete graph for a 3-hour journey.

**e.** Use the graph, and check by calculation, the distance travelled in:
45 minutes, $1\frac{1}{4}$ hours, 132 minutes.

**4.** Use a stop-watch to time a clockwork car over a distance of 5 metres. Draw a graph showing the speed of the car. Do not forget your title and scale.

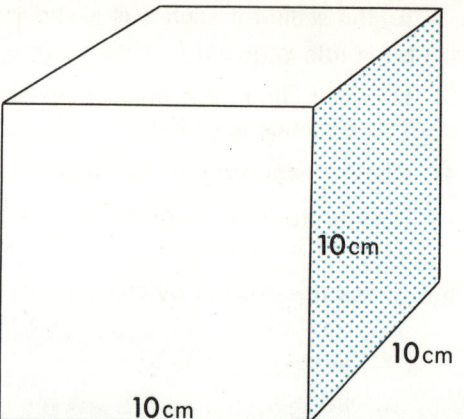

Here is a container which is 10 cm long,
10 cm wide, and 10 cm high.

1.  Write the missing numbers.

    a.  10× . . . = . . . centimetre cubes
        will fit on the bottom of the
        container.
    b.  The container will hold . . . layers
        of one-centimetre cubes.
    c.  The container will hold
        10×10× . . . = 1000 centimetre
        cubes.

●   A cube measuring 10 cm×10 cm×10 cm has a volume of
    1000 cubic centimetres (cm³) and a capacity of 1 litre.

    1000 cubic centimetres = 1 litre,
    and 1 millilitre (ml)      = 1 cubic centimetre.
2.  Make a cube which has a capacity of 1 litre or 1000 ml.
3.  Ask your teacher for a measuring jug
    which holds 1 litre and check that:

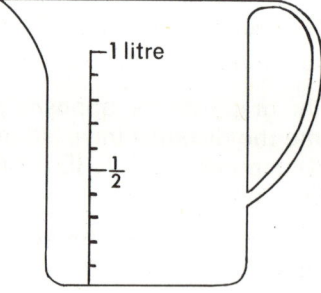

●   1 litre of water weighs 1 kilogramme (kg),
    or 1000 grammes (g).

4.  Do you think that 1 litre of milk
    weighs 1 kilogramme? Try it and see
    but do not forget that you want the
    **net** weight.
5.  Use the fact that 1 litre of water
    weighs 1 kilogramme to help you
    write the weight in grammes of 1
    cubic centimetre of water.

6.  Write the missing numbers in this working:  1 litre    = 1000 cm³
                                                 2 litres  = . . . cm³
                                                 0·5 litre = 500 cm³
                              so 2·5 litres = 2·5× . . . = 2500 cm³
                                                        or 2500 ml

●   To change litres to cubic centimetres or millilitres, **multiply** by 1000.

7.  Complete this working:  1000 cm³ = 1 litre
                            3000 cm³ = . . . litres
                             250 cm³ = 0·25 litre
        so 3250 cm³ = 3250 ÷ . . . = 3·25 litres

●   To change cubic centimetres or millilitres to litres **divide** by 1000.

1. Many wine bottles hold 1 litre of wine. Find an empty bottle and check whether or not it is a 1-litre bottle.

2. These bottles are full of sauce. What decimal of a litre is:

   a. bottle A?    b. bottle B?

   c. bottle A plus bottle B?

   d. Could bottle B be used to fill bottle A twice?

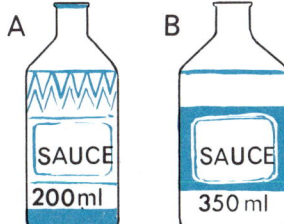

3. Fish tanks are supplied in various sizes. How many 10-cm cubes would fit into the following tanks?

   |   | Length | Breadth | Height |
   |---|--------|---------|--------|
   | a. | 10 cm | 20 cm | 10 cm |
   | b. | 100 cm | 10 cm | 10 cm |
   | c. | 100 cm | 100 cm | 10 cm |
   | d. | 1 m | 1 m | 1 m |

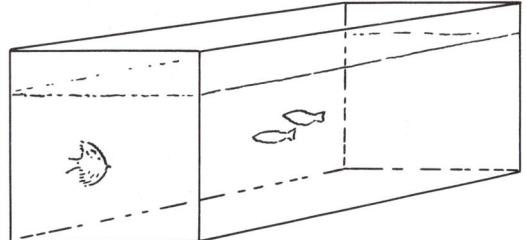

4. Change these capacities to litres:

   a. 100 cm³    b. 20 cm³
   c. 765 cm³    d. 1 cm³

5. Washo dish-washing liquid can be bought in 500-ml plastic bottles.

   a. Having used $\frac{1}{5}$ of the bottle, what decimal of a litre have I left?

   The weight of a full bottle is 750 g, and the empty bottle weighs 50 g.

   b. What fraction of the total weight is container?

   c. What fraction of the total weight is fluid?

   d. How much would the container full of water weigh? How could you tell if the container had been filled with water instead of washing-up liquid?

   e. How many containers could be filled by the manufacturer from a tank which holds 1000 litres?

**Averages**

Three sunflower seeds were planted, and they grew into fine plants. Here is a graph showing their heights.

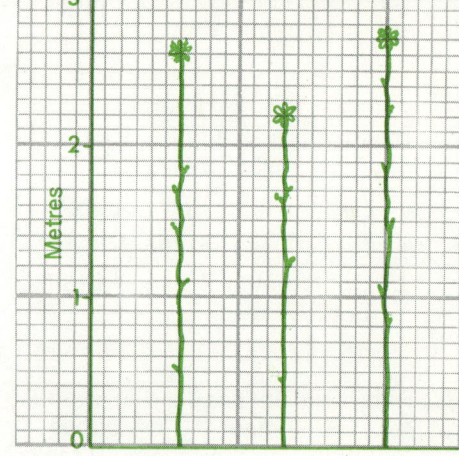

1. **a.** How tall is the tallest flower in: metres; centimetres?

   **b.** How tall is the shortest flower in: metres; centimetres?

   **c.** Add these three heights:
   2·7 m      2·3 m      2·8 m

   **d.** Cut a piece of string 780 cm long. This is as long as the three plants placed end to end.

   **e.** Write the length of $\frac{1}{3}$ of the piece of string. Check by folding the string into three parts.

Three plants, 2·6 metres tall, would have a total height of 7·8 metres, which is the total height of the three plants shown on the graph.

● 2·6 metres is called the **average** height of the plants.

To find the average of three heights:

A  add the three heights:        B  divide by 3:
   2·7 metres
   2·3 metres
   2·8 metres                          2·6
   ———————                          3)7·8
   7·8 metres

Average height = 2·6 metres

Here are four boxes of matches.
On each box is written
**Average contents: 48**
The flags show the actual number of matches in each box.

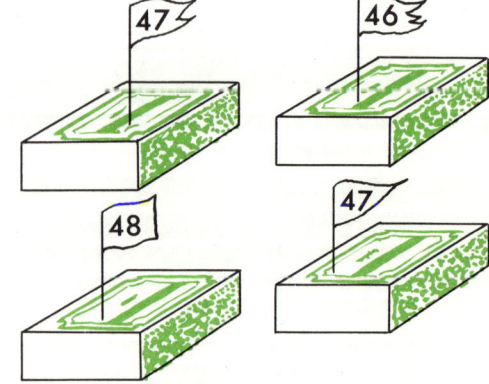

2. **a.** Find the total number of matches in the four boxes.

   **b.** Divide your total by 4.

The **average** number of matches in the four boxes is 47.

   **c.** Is 48+46+47+47 = 4×47?

If you tested the average contents of more boxes you would probably find that for a large number of match-boxes the **average contents** is 48.

1.  Here is a table showing the number
    of eggs laid by hen A and hen B.

| Day | 1 | 2 | 3 | 4 | 5 | 6 | 7 |
|-----|---|---|---|---|---|---|---|
| A   | 2 | 1 | 1 | 2 | 2 | 1 | 2 |
| B   | 1 | 2 | 0 | 1 | 0 | 1 | 2 |

    How many eggs were laid by:

    a.  hen A        b.  hen B
    c.  Write the average number of eggs laid per day by hen B.
    d.  What was the total number of eggs laid by both hens?

2.  This table shows the runs
    scored by three batsmen.

    a.  Complete the table.

    b.  Who has the highest
        average score?

    c.  Who has the lowest
        average score?

    d.  Find the total number of
        runs scored, and the **overall**
        average per innings for each batsman.

| | | Batsman | | |
|---------|---------|---------|---------|--------|
| Innings | Blaster | Tickler | Waller | |
| 1 | 20 | 10 | 5 | |
| 2 | 1 | 8 | 6 | |
| 3 | 3 | 9 | 4 | |
| 4 | 6 | 4 | 8 | |
| **Total** | 30 | | | |
| **Average** | 7·5 | | | |

3.  This is a list of the daily takings in the school tuck-shop:

| Monday | Tuesday | Wednesday | Thursday | Friday |
|--------|---------|-----------|----------|--------|
| £2·55  | £2·35   | £1·52     | £1·23    | £4·50  |

    Work out: a.  the total takings        b.  the average daily takings

4.  A farmer has 9 cows which produced 22680 litres of milk in 1 year. Work out the
    average:

    a.  monthly yield of milk        b.  monthly yield per cow

5.  A boy saved £1·35 in six weeks. What was his average weekly saving?

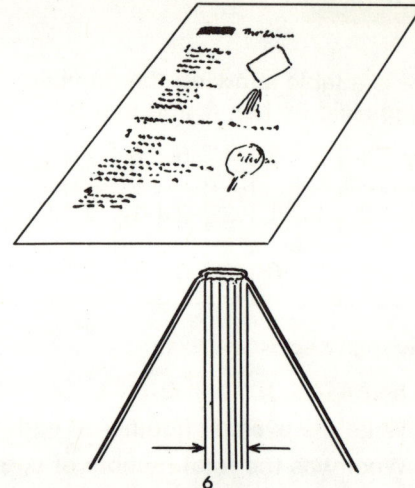

1. Measure the length and breadth of this page. How thick is the page? A ruler is not accurate enough to measure the thickness of one page.

2. a. Open this book until you have 50 sheets of paper. Take care, not 50 pages.

   b. Measure as accurately as you can the thickness of your 50 sheets.

$\frac{6}{10}$ cm

I measured my 50 sheets and found the thickness to be $\frac{6}{10}$ cm, or 0·6 cm.

Since 50 sheets are 0·6 cm thick
      5 sheets are 0·6 ÷10 = 0·06 cm thick
and   1 sheet is   0·06÷ 5 = 0·012 cm thick

3. Use the working above to help you write:

   a. 100 sheets are . . . cm×2 thick
                = . . . cm

   b. 1000 sheets are 0·6 cm×20 thick
                = . . . cm

Do you think that 1 sheet is **exactly** 0·012 cm thick? Remember that our measurements are only **approximate**.

A magnified picture of your ruler looks like this.

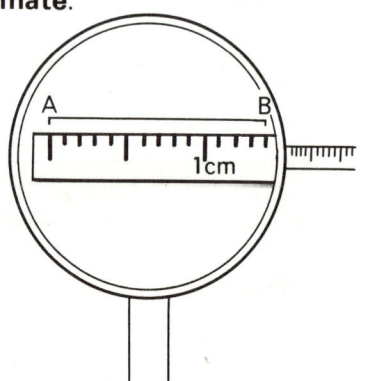

The line segment $\overline{AB}$ is $1\frac{4}{10}$ cm long or 1·4 cm.
Each centimetre is divided into 10 divisions which are called **millimetres** (mm).
**Remember** 10 millimetres equal 1 centimetre
              1 millimetre equals 0·1 centimetre

4. Complete this working and check your rule.

   a. 200 mm = 200÷10 cm
              = . . . cm

   b. 255 mm = . . . ÷ . . . cm
            = 25·5 cm

5. Check on your ruler that:
   12 cm = 12 × 10 mm
        = 120 mm

● To change centimetres to millimetres multiply by 10.

## Let's Try

1. Change these centimetres to millimetres:

   **a.** 10    **b.** 2·5    **c.** 1·25    **d.** 0·4    **e.** 0·35

2. Change these millimetres to centimetres:

   **a.** 20    **b.** 12    **c.** 105    **d.** 5    **e.** 9

3. Find the thickness of a page of your exercise book. It will help if you work on 100 sheets.
   Try estimating the thickness of other sheets of thin card or paper and measure their thickness.

4. A bobbin of machine cotton is 5 cm wide. By using a magnifying glass, I have counted 500 threads across the reel. Complete the following working.

   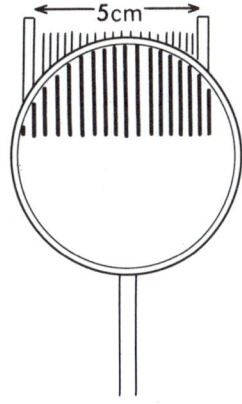

   **a.** The thickness of 1 thread
   $$= \frac{5}{500} \text{cm} = \ldots \text{cm}$$

   **b.** The thickness of 1 thread
   $$= \frac{?}{500} \text{mm} = \ldots \text{mm}$$

   **c.** The width of 100 threads
   $$= \ldots \text{cm} \times 100$$
   $$= \ldots \text{cm}$$

5. Use the following method of finding the thickness of a piece of wire or string:

   Start at 0 on your ruler and carefully wrap the string round it until you have covered 10 cm.
   Count the number of threads and then find:

   Thickness of string $= \dfrac{?}{10}$ cm $= \ldots$ cm

   or $= \dfrac{?}{100}$ mm $= \ldots$ mm.

   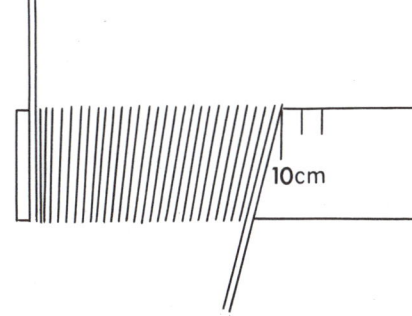

Here are 5 coins which are in daily use.

1. Write the missing numbers.
   a. . . . of A equals 1 of B
   b. . . . of C equals 1 of E
   c. . . . of C equals 2 of D

   A         B

2. Write the missing **fraction**.
   a. 5p is . . . of 10p
   b. 2p is $\frac{?}{5}$ of 5p
   c. $\frac{?}{5}$ of 10p = 2p
   d. $\frac{?}{20}$ of 10p = $\frac{1}{2}$p

C        D        E

- To write a sum of money as a fraction of £1 work in pence.

For example, to write 28p as a fraction of £1:

$$\frac{28p}{100p} = \frac{\overset{7}{28}}{\underset{25}{100}} \quad \text{or} \quad \frac{28p}{100p} = 0 \cdot 28$$

28p is $\frac{7}{25}$, or 0·28 of a £.

- To find a fraction of £1 change to pence.

For example, $\frac{3}{5}$ of £1 $= \frac{3}{5} \times 100p = \frac{3}{\underset{1}{5}} \times \overset{20}{100}p = 60p$

or $\frac{3}{5}$ of £1 $= 0 \cdot 6 \times £1 = 0 \cdot 6 \times 100p = 60p$

This is the sixth coin which we use:

3. Complete the following sentences:

   a. 50p is $\frac{?}{2}$ of £1.

   b. $\frac{1}{5}$ of 50p = . . . p

Peter is saving up for a new yacht which costs £2·50. So far he has saved 20p, 1½p, 4½p, 1½p, 50p, 12½p.

4. Complete this working for Peter:

   a. I have saved
   20
   1½
   4½
   1½
   50
   12½
   2
   $\left[\frac{1}{2}+\frac{1}{2}+\frac{1}{2}+\frac{1}{2}\right. = 2$

   Total ____ p

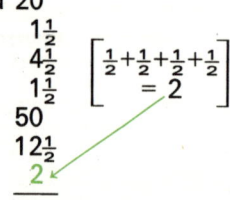

£2·50

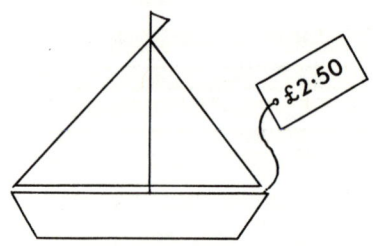

**b.** The fraction I have saved so far = $\dfrac{£0{\cdot}90}{£2{\cdot}50} = \dfrac{90}{250} = \ldots$

**c.** When Peter has saved $\frac{3}{4}$ of the money he needs he will have
$\frac{3}{4}$ of £2·50 = $\frac{3}{4} \times 250\text{p} = \ldots \frac{1}{2}\text{p}$

**d.** Peter will need to save another $62\frac{1}{2}\text{p}$ when he has saved $\frac{3}{4}$ of the cost.
Add $62\frac{1}{2}\text{p}$ to your answer to part **c** and check that the total is £2·50.

Let's Try

**1.** Write these pence as a **decimal** of a £.

    **a.** 2p     **b.** 15p     **c.** 40p     **d.** 120p

    **e.** 102p    **f.** 255p   **g.** 199p   **h.** 919p

**2.** Write these pence as a fraction of £2. Show your fraction in its lowest terms.

    **a.** 2p     **b.** 30p     **c.** 95p     **d.** 100p

    **e.** 105p    **f.** 150p   **g.** 175p   **h.** 199p

**3.** Work out these fractions of £1:

    **a.** $\frac{3}{10}$     **b.** $\frac{3}{5}$     **c.** $\frac{7}{20}$     **d.** $\frac{7}{40}$

**4.** Work out these fractions of £2:

    **a.** $\frac{7}{10}$     **b.** $\frac{3}{5}$     **c.** $\frac{19}{20}$     **d.** $\frac{11}{40}$

**5.** Jane is trying to save £3·50 for this
doll. Complete this table for Jane.

|  | Fraction saved | Amount saved | Amount to save |
|---|---|---|---|
| **a.** | $\frac{1}{10}$ |  | £3·15 |
| **b.** | $\frac{1}{5}$ | 70p |  |
| **c.** | $\frac{1}{4}$ |  |  |
| **d.** | $\frac{7}{10}$ |  |  |

**6.** Andrew's money-box contains £2·85
in 5p and 10p pieces. There are as
many 5p pieces as 10p pieces.

    **a.** How many of each coin are there
    in the box?

What fraction of Andrew's money is in:

    **b.** 5p pieces     **c.** 10p pieces

**Multiplication and Division of Fractions**

1.  One-fifth of this set is shaded blue.
    If 2 more elements are shaded blue,
    what fraction will be shaded?

    Look carefully at this equation

    $\boxed{\frac{1}{5}\times 3 = \frac{3}{5}}$  It will remind you that:

● When multiplying a whole number by a fraction, or a fraction by a whole number,
   multiply the numerator of the fraction by the whole number.

This diagram shows a ruler which has
been magnified.

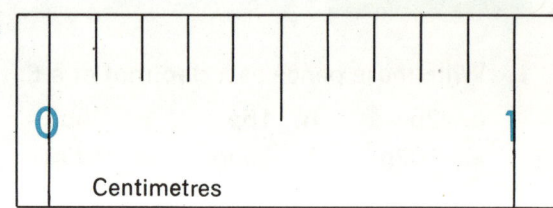

2.  **a.** Point to the mark which is $\frac{3}{5}$ cm
        from 0.
    **b.** Point to the mark which is only $\frac{1}{2}$
        as far along as the $\frac{3}{5}$ cm mark.

Centimetres

You should have found that $\frac{1}{2}$ of $\frac{3}{5} = \frac{3}{10}$

Check through this working:

**a.**  $\frac{1}{2}$ of $\frac{3}{5} = \frac{1}{2}\times\frac{3}{5}$      **b.** $\frac{3}{5}\times\frac{1}{2}$

$= \dfrac{1\times 3}{2\times 5} = \boxed{\dfrac{3}{10}}$      $= \dfrac{3\times 1}{5\times 2} = \boxed{\dfrac{3}{10}}$

● When multiplying one fraction by another, multiply the numerators and then
   multiply the denominators.

For example: $\frac{2}{5}$ of $\frac{3}{4} = \frac{2}{5}\times\frac{3}{4} = \frac{6}{20} = \frac{3}{10}$

3.  Check the example by drawing a
    pattern like this:
    Shade lightly $\frac{3}{4}$ of the pattern, and
    then darken $\frac{2}{5}$ of your shading.

4.  **a.** What is the whole number missing
        from this sentence?
        "To share between . . . we need to
        find $\frac{1}{2}$ of the box of marbles."
        $10\div\ldots = 10\times\frac{1}{2}$
    **b.** Is this a true statement?
        $9\div 3 = 9\times\frac{1}{3}$

● To divide by a whole number, n, multiply by $\dfrac{1}{n}$.

For example, $\frac{4}{5}\div 8 = \frac{4}{5}\times\frac{1}{8} = \frac{4}{40} = \frac{1}{10}$

1. Complete these divisions:

    **a.** $\frac{3}{4} \div 2$    **b.** $\frac{7}{10} \div 2$    **c.** $\frac{4}{5} \div 8$

    **d.** $\frac{5}{4} \div 10$    **e.** $1\frac{3}{5} \div 4 = \frac{8}{5} \div 4 = \ldots$    **f.** $2\frac{4}{5} \div 4$

2. Complete these multiplications:

    **a.** $\frac{3}{4} \times \frac{3}{5}$    **b.** $\frac{2}{5} \times \frac{3}{4}$    **c.** $\frac{3}{10} \times \frac{2}{5}$

    **d.** $\frac{7}{20} \times \frac{5}{14}$    **e.** $\frac{7}{10} \times \frac{3}{10}$    **f.** $4 \times \frac{2}{5}$

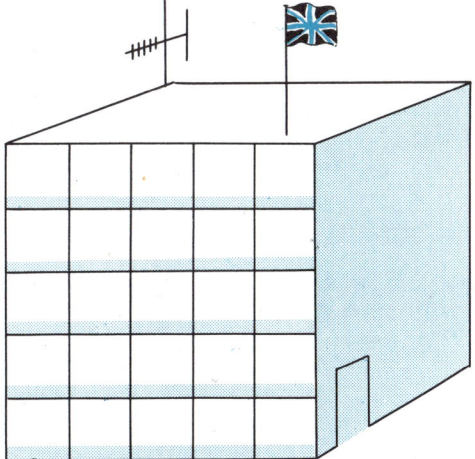

3. The window cleaner has started cleaning this side of a block of offices. On Monday he cleaned $\frac{2}{5}$ of the windows.

    **a.** How many windows were cleaned on Monday?

    **b.** What fraction is left to be cleaned?

    **c.** The window cleaner decides to clean $\frac{2}{3}$ of the rest of the windows on Tuesday and finish the job on Wednesday.
    How many will he clean on Tuesday and on Wednesday?

4. This is a plan of a kitchen floor.

    **a.** Copy the plan. Let 1 cm represent 1 metre.

    **b.** What fraction of the floor is shaded blue?

    **c.** Write the area of the blue portion in square metres (m²).

    **d.** Work out the area of a tile which measures $\frac{1}{4}$ metre by $\frac{1}{4}$ metre.

    **e.** How many tiles measuring $\frac{1}{4}$ m by $\frac{1}{4}$ m would be needed for the blue area and the grey area?

5 metres

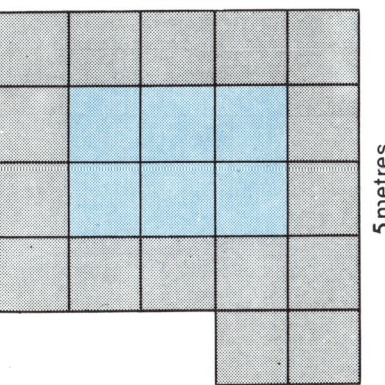

5 metres

5. Complete the following:

    **a.** $\frac{2}{3}$ of 1 right angle = . . . degrees

    **b.** $\frac{3}{5}$ of 1 kg = . . . g

    **c.** 1 km $\times \frac{7}{10}$ = . . . m

    **d.** $\frac{2}{5}$ of £1 = . . . p

Jane and Andrew have six 1p coins
between them.

1.  Complete these pairs which show
    Jane's share as the first number, and
    Andrew's share as the second number.
    (0, △), (6, △), (△, 4)

    This is a sentence which needs two
    replacements.

    $$\triangle + \square = 6$$

    Each replacement for Jane and Andrew
    is an element of

    {0, 1, 2, 3, 4, 5, 6}

2.  Complete this table for △+□ = 6

    | △ | 0 | 1 | 2 | 3 | 4 | 5 | 6 |
    |---|---|---|---|---|---|---|---|
    | □ |   |   | 4 |   |   |   |   |

    The relation R = {(0, 6), (1, 5), (2, 4), (3, 3), (4, 2), (5, 1), (6, 0)}
    It can be graphed:

    **a.** as an array                  or  **b.** as an arrow graph

    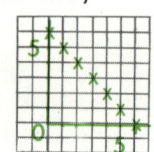
    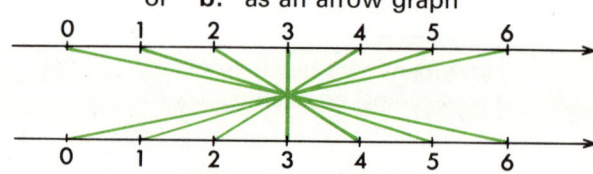

    R is the **solution set** to the sentence: △+□ = 6
    when △ and □ are replaced by elements from {0, 1, 2, 3, 4, 5, 6}
    This is another sentence with both △ and □ natural numbers.

    $$\triangle + \square < 10$$

3.  Check that the following pairs make the sentence true:

    (2, 7), (5, 4), (4, 4), (6, 3)

    Is △+□ **less than** 10 for each pair? Find some more pairs which make the
    sentence true.

4.  **a.**  Complete this set of pairs (△, □) which make △+□ < 8
        R = {(1, 1), (1, 2), . . . (2, 1), (2, 2), . . .}
    **b.**  Draw the array                          and the arrow graph for R

    Try using colour on your arrow graphs. For example,
    all the pairs (0, . . .) can be drawn in black,
    all the pairs (2, . . .) can be drawn in green, and . . .

1. Here is a set of toy soldiers. They can be shown marching in **rectangular** formations, for example,
   3 **ranks** (r) of 8 **soldiers** (s) = (3, 8)

   a. List the set of pairs which show all the formations r×s = 24.

   b. Draw an array which represents the solution set.

   c. Draw an arrow graph of the solution set on two parallel number lines.

2. a. A boy is strolling along at 3 kilometres per hour. Complete the following table.

   | Time taken | 1 | 2 | 3 | 4 | 5 | 6 | hours |
   |---|---|---|---|---|---|---|---|
   | Distance travelled | | | | | | 18 | kilometres |

   b. Draw an array of points from your table.

   c. Draw an arrow graph using two parallel number lines and the information from your table.

   d. How far should the boy travel in $2\frac{1}{2}$ hours? Show this on your array and on your arrow graph.

3. Farmer Giles wishes to fence in a sheep. He has 24 hurdles each 1 metre long.

   a. Complete this table showing the ways in which he can make a rectangular pen.

   | (L) Length | 1 | 2 | 3 | 4 | 5 | 6 | . . . metres |
   |---|---|---|---|---|---|---|---|
   | Breadth | 11 | 10 | | | | | . . . metres |
   | (A) Area | 11 | 20 | | | | | . . . metres |

   b. Draw an array of points, (L, A), from the table.

   c. Draw an arrow graph of the pairs (L, A).

   d. Which arrangement of hurdles will give the sheep most grass?

4. Write the solution set of pairs (△, □) where △ and □ are elements of {0, 1, 2, 3, 4, 5, 6, 7, 8, 9, 10}

   a. △ = 2□     b. △ = 10−□     c. $\frac{1}{3}$△ = □

   Draw arrays and arrow graphs of your solution sets.

**1.** Make two strips of card like this one and number them in **base two**.

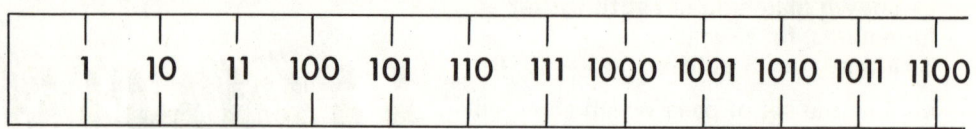

| 1 | 10 | 11 | 100 | 101 | 110 | 111 | 1000 | 1001 | 1010 | 1011 | 1100 |

If your strips are 30 cm long with 2 cm divisions you can number up to $1111_{(2)}$. This diagram shows how to use your cards for addition and subtraction of binary numbers. Keep the lower one still and slide the upper strip.

$100_{(2)} + 11_{(2)} = 111_{(2)}$

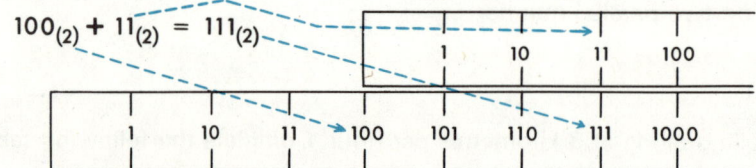

**2.** Slide your strip along to show $100_{(2)} + 11_{(2)} = 111_{(2)}$.
Check this addition by changing the numbers to base ten.

Remember that in binary arithmetic:

$$\begin{array}{r} 1 \\ +1 \\ \hline 10_{(2)} \end{array} \qquad \begin{array}{r} 10 \\ +\ 1 \\ \hline 11_{(2)} \end{array} \qquad \begin{array}{r} 11 \\ +\ 1 \\ \hline 100_{(2)} \end{array}$$

Check: $10_{(2)} = 2_{(10)}$    Check: $11_{(2)} = 3_{(10)}$    Check: $100_{(2)} = 4_{(10)}$

**3.** Follow through this working and then check by changing all the numbers to base ten.

| a. **Base two** | b. **Base two** | c. **Base two** |
|---|---|---|

$$\begin{array}{r} 100 \\ +101 \\ \hline 1001 \end{array} \qquad\qquad \begin{array}{r} 101 \\ +101 \\ \hline 1010 \end{array} \qquad\qquad \begin{array}{r} 111 \\ +101 \\ \hline 1100 \end{array}$$

Your strips of card can be used for subtraction. Set them in this position.

| 1 | 10 | 11 | 100 | 101 | 110 | 111 | 1000 | 1001 |

| 1 | 10 | 11 | 100 | 101 | 110 |

$111_{(2)} - 100_{(2)} = 11_{(2)}$

**4.** Check that $111_{(2)} - 100_{(2)} = 11_{(2)}$ by changing the numbers to base ten. When subtracting in binary remember that:

since $10_{(2)} + 1 = 11_{(2)}$, then $11_{(2)} - 1 = 10_{(2)}$   or   $\begin{array}{r} 11 \\ -\ 1 \\ \hline 10_{(2)} \end{array}$

and since $1 + 1 = 10_{(2)}$, then $10_{(2)} - 1 = 1$   or   $\begin{array}{r} 10 \\ -\ 1 \\ \hline 1_{(2)} \end{array}$

**5.** Follow through this working and then check by changing all the numbers to base ten

| a. **Base two** | b. **Base two** | c. **Base two** |
|---|---|---|
| 1101 | 1101 | 110 |
| $-$ 101 | $-$ 11 | $-$ 11 |
| 1000 | 1010 | 11 |

Let's Try

**1.** Use your strips to help you with these additions. Do not forget to check by changing to base ten.

   **a.** $10_{(2)}+10_{(2)}$    **b.** $11_{(2)}+1001_{(2)}$    **c.** $110_{(2)}+110_{(2)}$

**2.** Add the binary numbers and then check in base ten.

| a. | b. | c. |
|---|---|---|
| 10101 | 10101 | 10111 |
| + 1010 | +10001 | +10010 |
| _____ (2) | _____ (2) | _____ (2) |

   **d.** $11110_{(2)}+1010_{(2)}$    **e.** $10101_{(2)}+10101_{(2)}$    **f.** $10110_{(2)}+110101_{(2)}$

**3.** Work these subtractions on your strips. Check your answers.

   **a.** $101_{(2)}-11_{(2)}$    **b.** $110_{(2)}-101_{(2)}$    **c.** $1000_{(2)}-101_{(2)}$

**4.** Subtract the binary numbers and then check in base ten.

| a. | b. | c. |
|---|---|---|
| 10100 | 10101 | 10000 |
| $-$ 1100 | $-$ 1011 | $-$ 111 |
| _____ (2) | _____ (2) | _____ (2) |

   **d.** $10000_{(2)}-1111_{(2)}$    **e.** $11011_{(2)}-1110_{(2)}$    **f.** $111011_{(2)}-10101_{(2)}$

**5.** Write the binary numbers shown on the punched tape and then find their **sum** and **difference**.

 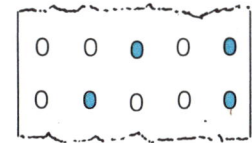 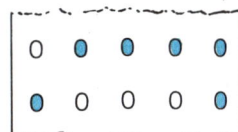

**6.** Write the number of days in July, August, and September in binary numbers and then find the total number of days.

**7.** Find out more about computers. Make a collection of pictures, tapes, and mechanical aids such as "slide rules".
You might even be able to arrange a visit to see a computer.

**The Load**

Have you seen **containers** like this
one which is being lifted from a truck
onto a train?

This container will carry a load of
40 tonnes.
1 **tonne** = 1000 **kilogrammes**

1. Write the missing numbers.

   **a.** 40 tonnes = . . . kilogrammes
   **b.** 40000 kilogrammes
      = 40000 ÷ . . . tonnes
      = 40 tonnes

● To change **tonnes** to **kilogrammes**
   multiply by 1000. To change
   **kilogrammes** to **tonnes** divide by
   1000.

The **tare weight** of a **container** carrying its maximum load is 50 tonnes, that is,
the weight of the container plus its load is 50 tonnes.

2. Complete these fractions:

   **a.** The fraction of the tare weight which is load is $\dfrac{40 \text{ tonnes}}{50 \text{ tonnes}} = \dfrac{?}{?}$

   **b.** The fraction of the tare weight which is container is $\dfrac{?}{?} = \dfrac{1}{5}$

Sometimes a container load might be shared by two customers. For example,
Mr. Smith uses $\frac{2}{5}$ of a container load and the rest is used by Mr. Brown. The charge
for a full load is £60.

3. **a.** Complete this working for Mr. Smith.
      Mr. Smith's share of the weight is $\frac{2}{5}$ of 40 tonnes = . . . tonnes
      His share of the cost is $\frac{2}{5}$ of £60 = £ . . .
   **b.** What fraction of the load is used by Mr. Brown?
   **c.** Write the missing numbers for Mr. Brown.

      Mr. Brown's share of the weight is $\dfrac{?}{?}$ of . . . tonnes = 24 tonnes.

      Mr. Brown's share of the cost is $\dfrac{?}{?}$ of £ . . . = £36.

Since Mr. Smith and Mr. Brown have shared the full load:
(Smith's payment) + (Brown's payment) = £60
and
(Smith's part of the weight) + (Brown's part of the weight) = 40 tonnes.

1.  Change these tonnes to kilogrammes:

    **a.** 10     **b.** 9·2     **c.** 0·75     **d.** 0·08

2.  Write these kilogrammes as tonnes:

    **a.** 2000     **b.** 210     **c.** 907     **d.** 50

3.  Complete the following:

    **a.** $\frac{3}{4}$ of 72 tonnes = . . . tonnes     **b.** $\frac{4}{5}$ of 21 tonnes = . . . tonnes
    **c.** $\frac{1}{5}$ of 3 tonnes = . . . kg     **d.** $\frac{3}{10}$ of 3 tonnes = . . . kg

4.  Write the first weight as a fraction of the second weight.

    **a.** 2 tonnes, 4 tonnes     **b.** 500 kg, 2 tonnes
    **c.** 800 kg, 3 tonnes     **d.** 124 kg, 2 tonnes

5.  This is an advertisement for a car
    with details of its weights.

    **a.** What is the weight of fuel, oil,
    and water?

    What fraction of the kerb weight is:
    **b.** dry weight
    **c.** fuel, oil, and water
    **d.** Write the two weights as decimals
    of a tonne.

    | Weight |
    | --- |
    | Dry 920 kg |
    | Kerb (including fuel, |
    | oil and water) |
    | 1050 kg |

6.  This truck will carry a load of 20
    tonnes.

    **a.** If it already contains $\frac{3}{5}$ of its load:
    how many tonnes have been
    loaded?
    how many tonnes are still to be
    loaded?
    **b.** If 12·5 tonnes have been loaded:
    what fraction of the loading has
    been done?
    how many kilogrammes are there
    still to load?

    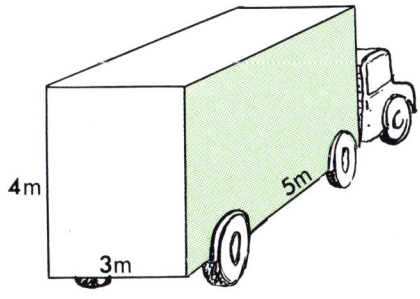

    4m     5m     3m

    The inside measurements of the truck are shown on the diagram.

    **c.** Work out the volume of the truck.
    **d.** Will 60 crates, each 1-metre cube, fit into the truck?
    Can the truck carry 60 of these crates if each crate weighs:
    300 kg; 400 kg?

**Polyhedra**

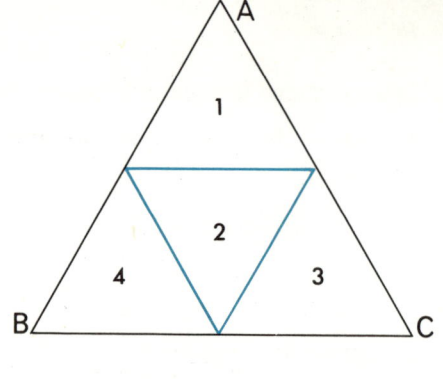

1. **a.** Copy this network onto stiff
   paper. Each triangle is an
   equilateral triangle.

   **b.** Carefully cut out the shape and
   fold along the blue lines. The three
   vertices A, B, and C should meet.

   **c.** Use sticky tape to fasten your
   model together.

   You have made a mathematical solid
   which has four faces and is a pyramid.

2. **a.** How many edges has your solid?

   **b.** How many vertices has your solid?

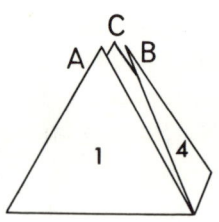

● A pyramid with four faces is called a
  **tetrahedron**.

3. Stick two tetrahedra together.
   Make sure that they are the same size.
   Write the missing numbers in
   the following sentence.
   My solid has 6 faces, . . . vertices,
   and . . . edges.

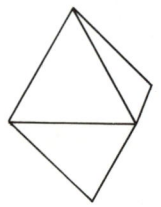

● A mathematical solid which has flat
  or plane faces is called a **polyhedron**.

   Here are 5 regular **polyhedra**
   (plural for polyhedron).

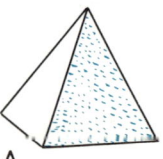
A

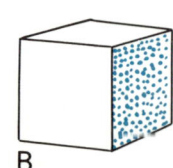
B

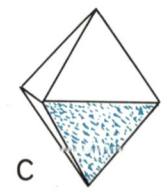
C

D

E

4. Can you see why they are called
   **regular** polyhedra?

   Here is a net for a square-based pyramid.

5. **a.** Draw the net and build the pyramid
   by folding along the blue lines.

   **b.** Fill in the missing numbers.
   A square-based pyramid has 5
   faces, . . . edges, and . . . vertices.

1. Which of the five regular solids on the opposite page have been constructed from:

   a. squares    b. equilateral triangles

2. This is a net for the third regular solid. It has **eight faces** and so is called an **octahedron**.

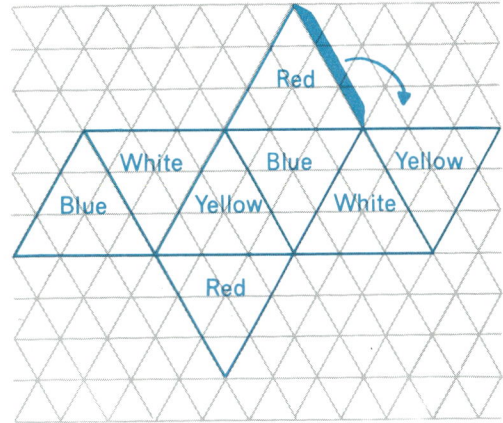

   a. Draw and cut out the net. You may put flaps on the net and stick with glue, or use sticky tape.

   b. How many vertices has your octahedron?

   c. How many edges has your octahedron?

   d. Paint the faces using the colours shown on the net and then hang your model on a piece of thread.

3. Here is a net for the fifth regular solid. This polyhedron is called an **icosahedron**.

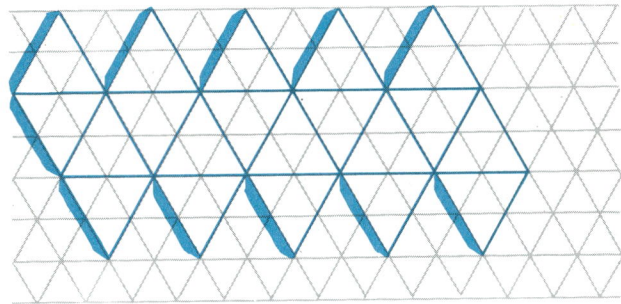

   a. Build an icosahedron using this net and fill in the missing numbers in the following sentence.
   My icosahedron has . . . faces, . . . edges, and . . . vertices.

   b. Turn your icosahedron over several times. Does it always look the same? It should if it is **regular**.

4. This picture shows an interesting solid which can be made by sticking 6 square-based pyramids onto a cube. Make a cube using a 5-cm square. Make the base of each pyramid a 5-cm square and build this solid.

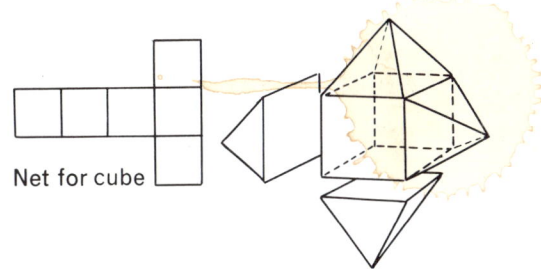

Net for cube

5. In question **4** you were given the net for a cube. Draw some different nets for a cube.

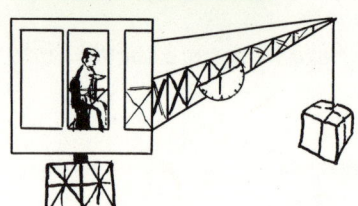

## Let's Discover

## How to Measure Angles

On the jib of a crane, we often see an angle measurer like this. The crane driver uses it to measure the angle at which the jib is resting.

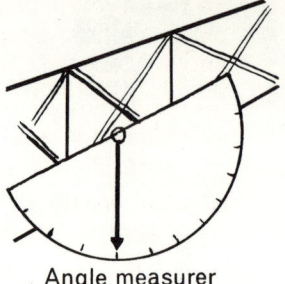

Angle measurer

1. Through how many right angles has the minute hand turned in each of these four pictures?

   In the fourth picture the hand has moved through 1 **revolution**.

2. Complete these sentences for a 12-hour clock:

   a. In one day the minute hand turns through t . . . revolutions.

   b. In 36 hours the minute hand turns through . . . revolutions.

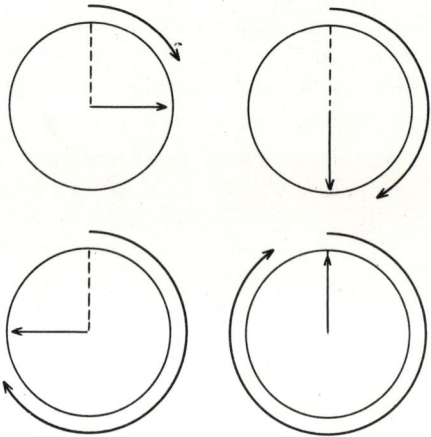

---

● In order to measure angles, one revolution is divided into 360 units. Each unit is called a **degree**. The short form for 1 degree is 1°.

---

3. a. How many right angles make a full revolution?

   **1 right angle = 90°.**

   b. How many right angles make $\frac{1}{2}$ a revolution?

   $\frac{1}{2}$ **a revolution = 180°.**

4. a. Draw and cut out a semicircle of tracing paper and trace this angle measurer.

   b. Complete the following sentences:

   On the diagram each right angle is divided into . . . equal parts.
   $90° \div 9 = \ldots°.$

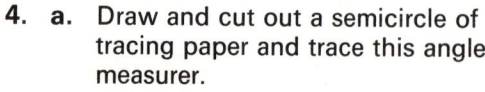

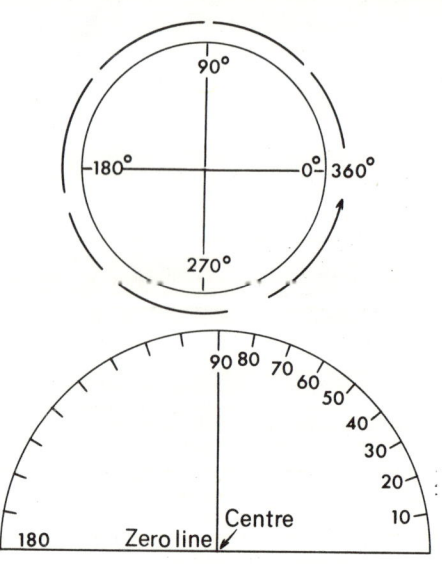

---

● You have made yourself an angle measurer which is called a **protractor**.

**5.** To measure the angle ATB, place the protractor on the angle. Make sure that:

   **a.** the centre of your protractor is on the vertex T; and

   **b.** the zero line is along one arm of the angle.

   m(∠ ATB) = 60°

**6.** Try using your protractor to measure angle PTR.
m(△ PTR) = 120°.

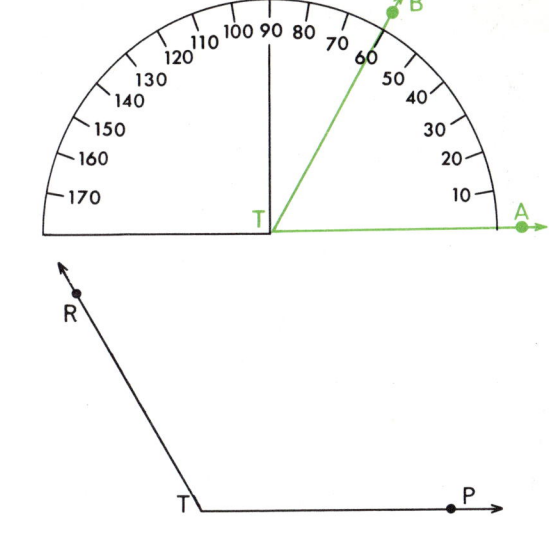

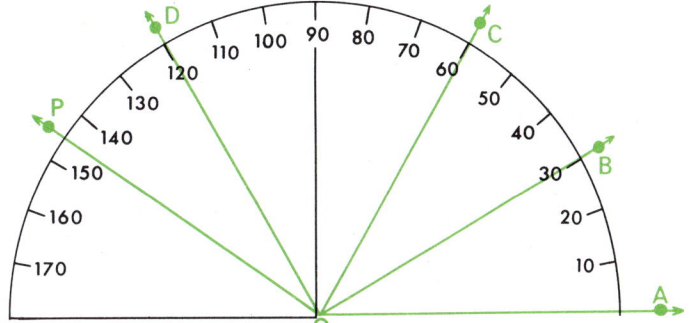

**1.** Complete the measure of each angle shown on this protractor.

   **a.** m(∠ AOB) = . . .

   **b.** m(∠ AOC) = . . .

   **c.** m(∠ AOD) = . . .

   **d.** m(∠ AOP) = . . .

   **e.** m(∠ BOC) = . . .

   **f.** m(∠ COD) = . . .

   **g.** m(∠ BOD) = . . .

   **h.** m(∠ COA) = . . .

**2.** Use your protractor to measure the angle between the hands of this clock at:

   **a.** 1 o'clock    **b.** 2 o'clock

   **c.** 3 o'clock    **d.** 4 o'clock

   **e.** 5 o'clock    **f.** 6 o'clock

**3. a.** Draw an angle which you think is about 70°. Check the measure of the angle with your protractor.

   **b.** Draw an angle which you think is about 130°. Check it with your protractor.

**4. a.** Open a door to an angle of 90°. Guess the angle, then ask a friend to check it with his protractor.

   **b.** Repeat part **a** with an angle of 30°, and then an angle of 100°.

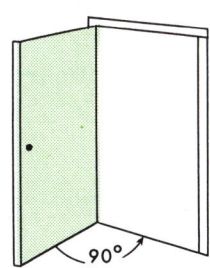

This is a triangle which divides the
plane into two regions.
The two regions are **interior** (inside),
            and **exterior** (outside).
The three interior angles are marked
A, B, and C.

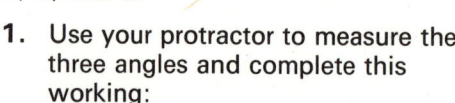

1. Use your protractor to measure the
   three angles and complete this
   working:

   m(∠ A) = . . .°
   m(∠ B) = . . .°
   m(∠ C) = . . .°
   **Total   180°**

The three angles of this triangle add
up to ½ a revolution or 180°.

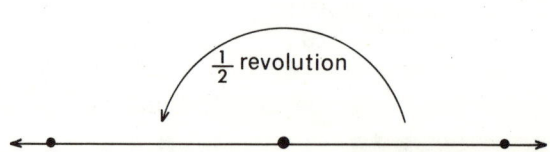

2. a. Draw and cut out a triangle like
      this one.
   b. Colour each interior angle with a
      different colour.
   c. Tear off each angle.
   d. Fit the three angles snugly
      together and stick them in your
      book.
   e. Copy and complete this sentence.
      The three angles make . . .
      revolution.

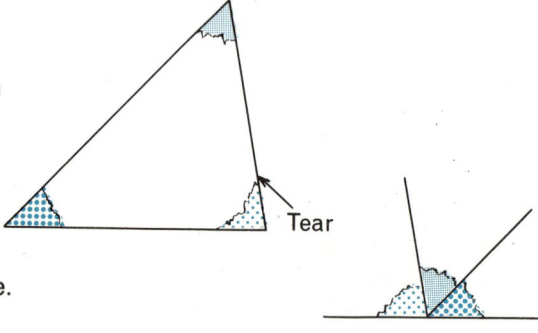

Tear

● The sum of the interior angles is 180°.

Here is a quadrilateral.

3. a. This quadrilateral is called a r . . .
   b. Colour the four interior angles.
   c. Tear off the angles and stick them
      snugly together on a piece of paper.

4. a. Now draw and cut out a
      quadrilateral like this one.
   b. Tear off the angles and fit them
      together.

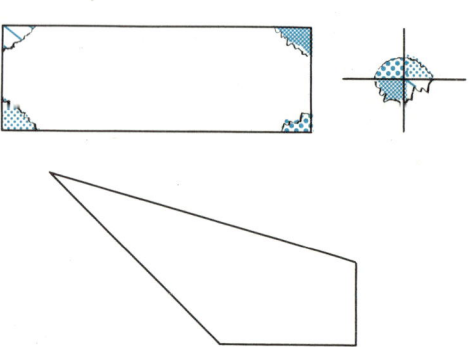

● The sum of the interior angles of a quadrilateral is 360°, or 4 right angles, or 1
  revolution.

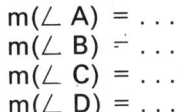

Let's Try

1. This quadrilateral is called a rhombus.

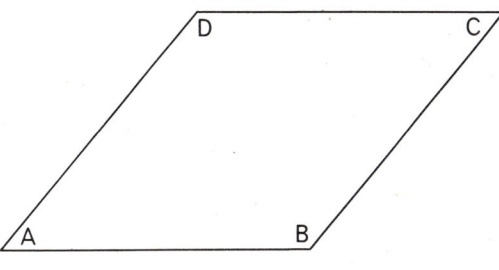

   **a.** Measure each of the angles and complete this working:

      m(∠ A) = . . .
      m(∠ B) = . . .
      m(∠ C) = . . .
      m(∠ D) = . . .
     **Total**

   **b.** Which two angles are equal, and less than a right angle?

   **c.** Which two angles are equal, and greater than a right angle?

2. Trace and cut out these quadrilaterals and check the sum of the interior angles by colouring and tearing off the angles.

   **a.**           **b.**           **c.**           **d.**

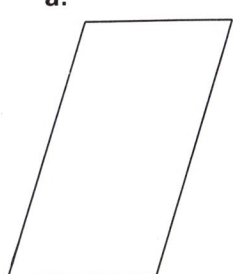 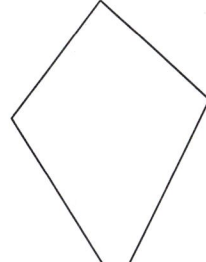 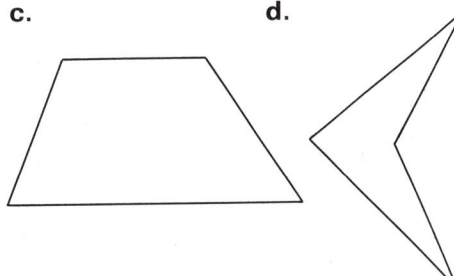

3. This is a picture of a farm gate. Look what happens when the cross piece is removed.

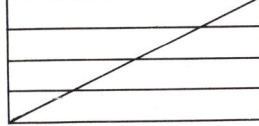 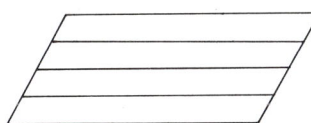

   **a.** Point to the interior angles of the large rectangle of the first gate.

   **b.** What is the measure of each interior angle of the large rectangle of the first gate?

   **c.** What is the measure of the same angles on the second gate?

   **d.** Does the sum of the angles you measured change as the gate collapses?

4. **a.** Draw and cut out a **regular** hexagon.

   **b.** Tear off and fit three angles together.

   **c.** What is the sum of the 6 angles of a hexagon?

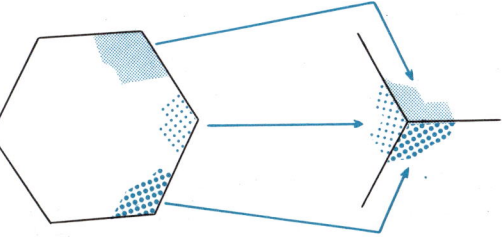

Here is a picture of a 5-minute timer.
Start the clock at 1, and 5 minutes
later it will show 1 again.

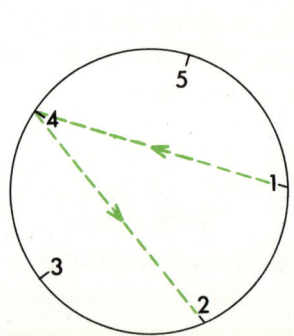

1.  **a.**  Starting at 5, what time will the
    clock show after:

| | 1 | 4 | 7 | 10 | 13 | 16 | minutes |
|---|---|---|---|---|---|---|---|
| Finish | 1 | 4 | | 5 | | | |

**b.** Copy the clock-face and join up
the finishing times from your table
in the same order, that is 1 → 4 → . . .

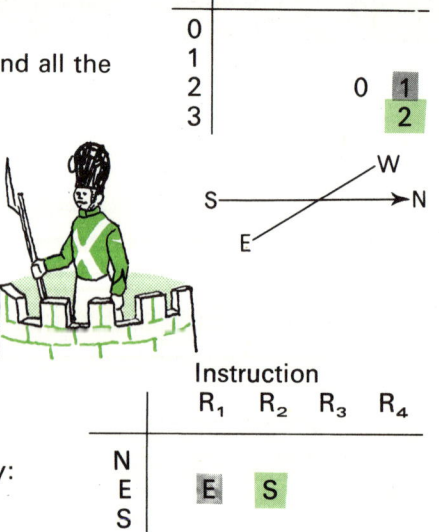

● You have been working in modulo 5 arithmetic.

For example, $17 \div 5 = 3$ rem. 2      so $17 = 2$ (mod. 5)

but $17 \div 7 = 2$ rem. 3      so $17 = 3$ (mod. 7)

In modulo 5 any number can be replaced by an element of {0, 1, 2, 3, 4}.

● A modular system is called a **finite arithmetic**.

2.  Complete this addition table for the finite arithmetic
    mod. 4:
    For example, $2+2 = 4 = 0$ (mod. 4)
    Shade all the 1s in black, all the 2s in green, and all the
    3s in blue.
    This picture shows a sentry on duty.
    The instruction $R_1$ says turn   360°
                   $R_2$ says turn   90°
                   $R_3$ says turn   180°
                   $R_4$ says turn   270°
    When facing East, the sentry is ordered "$R_2$".
    He turns through 1 right angle   and
    faces South.
    When facing East the sentry is ordered "$R_1$".
    He turns through 4 right angles and
    faces East.

| + | 0 | 1 | 2 | 3 |
|---|---|---|---|---|
| 0 | | | | |
| 1 | | | | |
| 2 | | | 0 | 1 |
| 3 | | | | 2 |

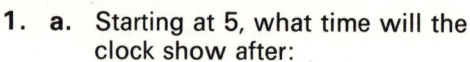

3.  **a.**  Copy and complete this table for the sentry:

    **b.**  In the table shade Es black, Ss green,
    and Ws in blue.

    Compare the sentry's table with your mod. 4 addition table.

| | Instruction | | | |
|---|---|---|---|---|
| | $R_1$ | $R_2$ | $R_3$ | $R_4$ |
| N | | | | |
| E | E | S | | |
| S | | | | |
| W | | | | |

● The two tables are said to have the same **structure** since they have the same
pattern. The sentry's table is another example of a **finite arithmetic**.

1. The teacher has a set of cards on which the natural numbers are printed. The cards are put into boxes labelled **Ian**, **Alan**, **Betty**.

   Ian (I) has numbers which are equal to 0 mod. 3.
   Alan (A) has numbers which are equal to 1 mod. 3.
   Betty (B) has numbers which are equal to 2 mod. 3.

   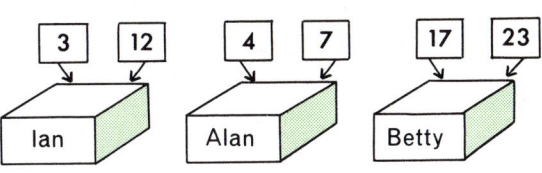

   **a.** What kind of number do you get if you add an "Alan number" to a "Betty number"?

   **b.** Complete the addition table for I, A, and B.

   | + | I | A | B |
   |---|---|---|---|
   | I |   |   |   |
   | A |   |   |   |
   | B |   |   |   |

2. **a.** Complete a mod. 4 multiplication table.

   **b.** Has this table the same structure as the sentry's table?

3. **a.** Try adding the rows and columns of this number square. It is a **magic square**.

   | 4 | 9 | 2 |
   |---|---|---|
   | 3 | 5 | 7 |
   | 8 | 1 | 6 |

   | × | 0 | 1 | 2 | 3 |
   |---|---|---|---|---|
   | 0 |   |   |   |   |
   | 1 |   |   |   |   |
   | 2 |   |   | 0 |   |
   | 3 |   |   |   | 1 |

   **b.** Change all the numbers to mod. 5. Is it still a magic square in mod. 5?

   **c.** Change all the numbers to mod. 4. Is it still a magic square in mod. 4?

4. Here is a square lid which just fits on a box.

   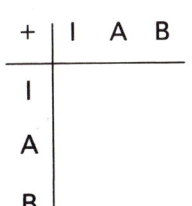

   **a.** In how many ways can it be fitted onto the box without turning it over?

   **b.** Starting in the position shown in the diagram,
   $R_1$ says "Turn 360°"
   $R_2$ says "Turn 90°"
   $R_3$ says "Turn 180°"
   $R_4$ says "Turn 270°"
   Draw the positions for $R_1$, $R_2$, $R_3$, and $R_4$.

   | + | $R_1$ | $R_2$ | $R_3$ | $R_4$ |
   |---|---|---|---|---|
   | $R_1$ |   |   |   |   |
   | $R_2$ |   |   |   |   |
   | $R_3$ |   |   | $R_4$ |   |
   | $R_4$ |   |   |   |   |

   **c.** $R_2+R_3$ means "do $R_3$ then $R_2$", so $R_2+R_3$ is turn 180° then 90°. Check that $R_2+R_3 = R_4$ and then complete this table.

   **d.** Does your table have the same structure as the mod. 4 addition table?

**Arrays and Relations**

1. **a.** Make two tetrahedra. The net for a tetrahedron is on page 50.

   **b.** Number the faces of each tetrahedron 1, 2, 3, and 4.

   **c.** The two solids can be used as dice. Complete this table showing all the ways in which the tetrahedron can fall. The numbers are for the "face on the table". For example: (1, 3) is one pair, (3, 1) is another.

   **d.** Complete this graph of **the set of ordered pairs** from your table:

**Table of possible pairs**

| | | | |
|---|---|---|---|
| (1, 1) | (2, 1) | (3, 1) | (4, 1) |
| (1, 2) | (2, 2) | (3, 2) | ( , ) |
| ( , ) | ( , ) | ( , ) | ( , ) |
| (1, 4) | ( , ) | ( , ) | ( , ) |

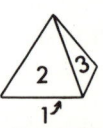 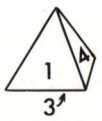

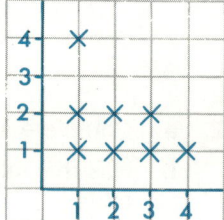

**Graph of possible pairs**

● Your graph of the set of pairs is called a **rectangular array**.

Here is another rectangular array.

2. Complete the following sets:

   **a.** A is the set which forms the array.
   A = {(1, 1), (1, 2), (2, . . .}

   **b.** P, a subset of A, contains elements for which the two parts add up to an **even** number.
   P = {(1, 1), (2, __), (__, __)}

   **c.** Q, a subset of A, contains elements for which the two parts are the same.
   Q = {(__, __), (__, __)}

   **d.** Check that these graphs are correct for:

**Graph of set A**

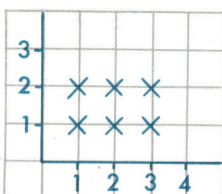

P                    Q                    P ∩ Q

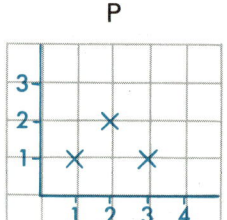    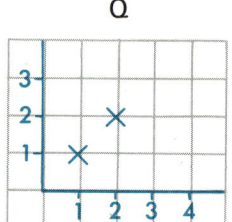    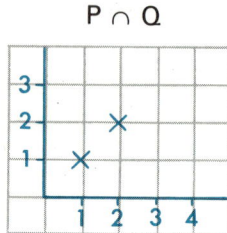

● The graphs of A, P, Q, and P ∩ Q are graphs of **relations** since a relation is a **set of ordered pairs**.

1. Copy this array and then graph and
   list the subset of pairs which can be
   joined to form:

   a. a square, for example
      {(1, 1), (1, 2), (2, 1), (2, 2}

   b. a rectangle but not a square

   c. a parallelogram

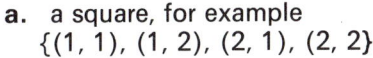

2. a. Make two similar cubes and
      number the faces 1, 2, 3, 4, 5,
      and 6. Using the cubes as dice,
      complete this table for all the
      possible ways in which the dice
      can fall.

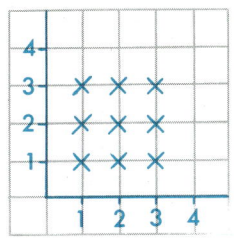

   | | Possible pairs | |
   |---|---|---|
   | (1, 1) | (2, 1) | ( , ) |
   | (1, 2) | (2, 2) | |

   b. Draw the rectangular array (A)
      which shows all the possible pairs.

   c. Draw a new pair of axes and graph
      these subsets of A:
      P, the relation which contains pairs
      such that the first element equals
      the second ($\triangle$, $\triangle$).

      Q, the relation of elements ($\triangle$, $\square$)
      such that $\triangle + \square = 7$

   d. Draw the graphs of $Q \cup P$ and $Q \cap P$

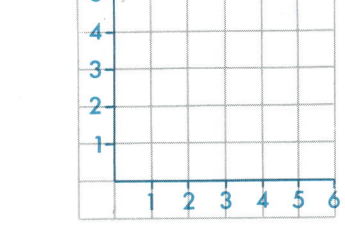

3. Here are three seats for Alan, Betty,
   and Chris. Each pupil can sit on any
   one of the seats. The array (1) shows
   all the possible seating positions.

   The graph (2) shows one actual
   seating arrangement,
   $R_1 = \{(A, 1), (B, 2), (C, 3)\}$

   a. Is $R_1$ a subset of A?

   b. Draw graphs of $R_2, R_3, \ldots$ which
      show all the possible seating
      arrangements.

   c. How many different seating
      arrangements are there?

   d. How many arrangements are there
      in which Alan sits next to Betty?

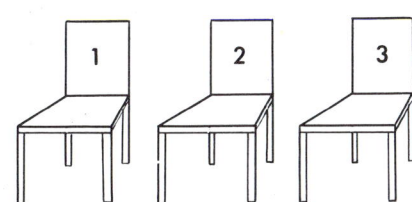

Graph of A          Graph of $R_1$

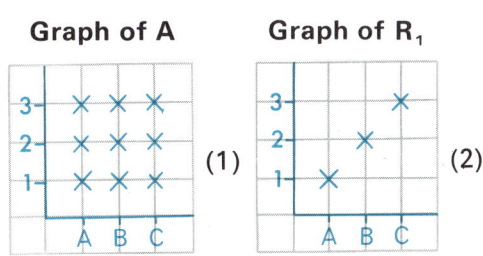

(1)          (2)

**Circumference**

1. Choose a container which is circular and check the distance round it by using the following methods:

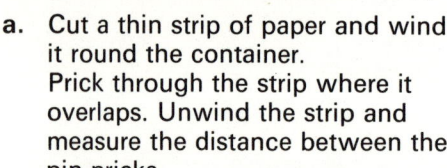

a. Cut a thin strip of paper and wind it round the container.
Prick through the strip where it overlaps. Unwind the strip and measure the distance between the pin pricks.

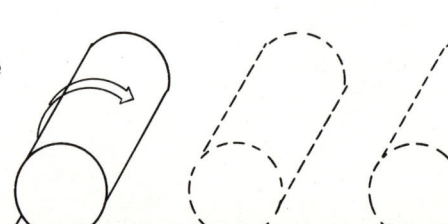

b. Make a mark on the edge of the container. Match the mark to a mark on the edge of a piece of paper.
Roll the container through a full turn. Measure the distance between the starting and finishing points on the paper.

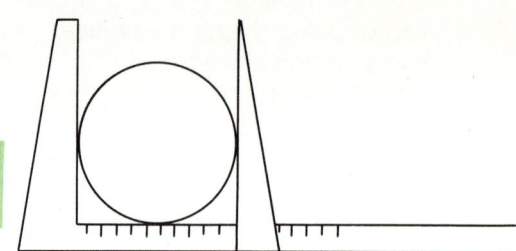

● The distance round a circle is called its **circumference**.

2. Measure the diameter of your container using a pair of calipers.

Here is a table showing the diameter and circumference of four circles.

| Circle | Diameter | Circumference |
|--------|----------|---------------|
| 1 | 5 cm | 15·7 cm |
| 2 | 10 cm | 31·4 cm |
| 3 | 12 cm | 37·9 cm |
| 4 | 15 cm | 47·1 cm |

A graph of the diameter and circumference of each circle has been started for you.

3. a. Draw and complete the graph with scale:
1 cm represents 5 cm of circumference
1 cm represents 1 cm of diameter
Extend the circumference axis to 50 cm.

b. Check the diameter and circumference of your own circle on the graph.

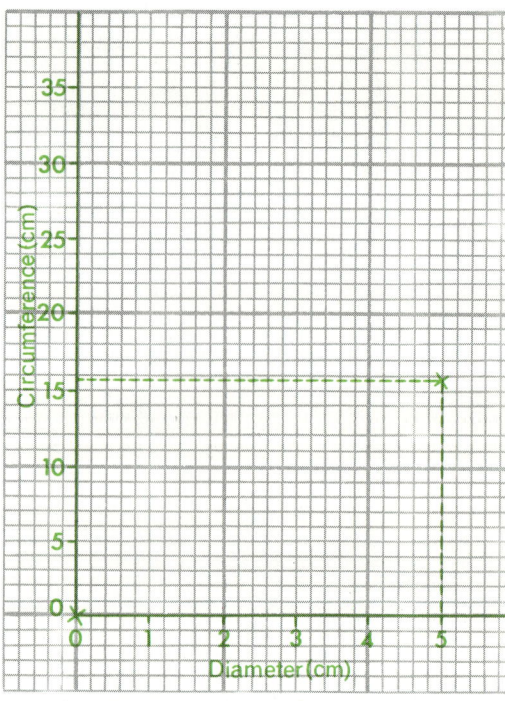

c. Divide the **circumference** of each circle by its **diameter**. For example:

Circle (1) $\dfrac{C}{D} = \dfrac{15\cdot7}{5} = 3\cdot14$

**d.** Is each answer to C÷D *about* 3·14?

For example, a circle with a diameter of 13 cm will have a circumference
= 3·14×13 cm
= 3·14×(10+3) cm
= 31·4+9·42 cm
= 40·82 cm

Check this on your graph.

**1. a.** Use a pair of compasses to draw a circle with a radius of 10 centimetres.
  **b.** Check the circumference of your circle. Try using a fine chain or cord laid round the edge of the circle.
  **c.** Divide your answer to part **b** by 20.

**2.** The school gardener wished to cut out a flower-bed from a lawn. He tried using a loop of string, two pegs, and a sharp stick.

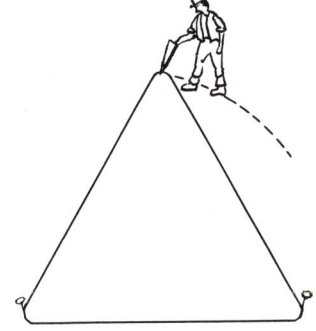

  **a.** Use paper, drawing pins, and a pencil to find out the shape of the bed.
  **b.** Move the pins nearer together and try again.
  **c.** Now try using only one pin. What is the shape of the flower-bed now?

**3.** Archimedes found that the circumference of a circle is $3\frac{1}{7}$ times the length of its diameter.

  **a.** Find out more about Archimedes.
  **b.** Change $3\frac{1}{7}$ to a decimal. Is $3\frac{1}{7}$ exactly equal to 3·14?

**4.** Work out the circumference of a circle with diameter:

  **a.** 40 cm    **b.** 25 cm    **c.** 18 cm    **d.** 1 metre

**5.** Find the circumference and diameter of some large circles such as a bicycle wheel.

**Some Properties of a Cylinder**

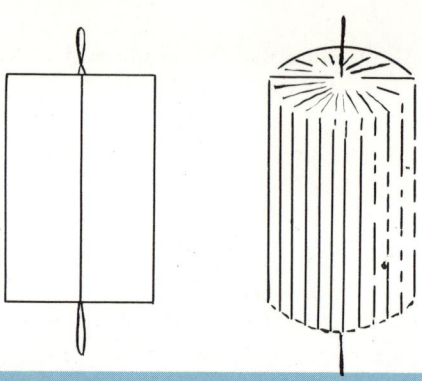

1. **a.** Cut out a rectangular piece of card. Draw a **line of symmetry** on it.

   **b.** Attach a rubber band near each end of the line of symmetry. Ask your neighbour to "wind up" the card and then let it spin.

   **c.** Did you get the impression of a cylinder as the card spun?

---

- The ends of a cylinder are usually circular regions.
- The height of a cylinder is the shortest distance between the circular ends.

---

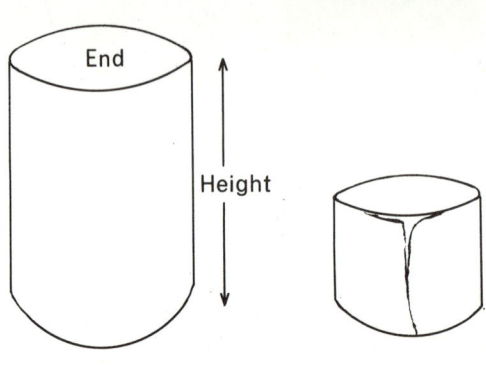

2. **a.** Wrap a sheet of paper round a cylinder. Trim the paper until its ends just meet and the curved surface is just covered.

   **b.** Open up your paper and answer these questions.
   What shape is your piece of paper?
   Which edge of the paper is the same length as the circumference of the end of your cylinder?
   Which edge of the paper is the same length as the height of your cylinder?

---

- The curved surface of a cylinder can be made from a rectangle.

---

A cylinder is a closed surface which can be made from this net.

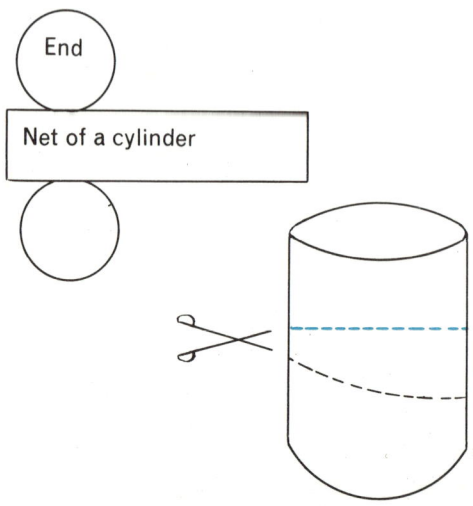

3. **a.** Make a cylinder by joining opposite edges of a rectangular sheet of paper.

   **b.** Press your cylinder flat as in this diagram. Make a cut along a line which is parallel to the blue line of symmetry.

   **c.** You should now have two cylinders if you open up the pieces. Are the ends still circular?

   **d.** Make a fresh cylinder. Press it flat and cut along the black line. Are the ends of your cylinders still circular?

- The line segment joining the centres of the circular ends of a cylinder is called the **axis**.

**4.** Look at a roller-skate wheel.  Is the wheel a cylinder? Does the axle form the axis of the cylinder?

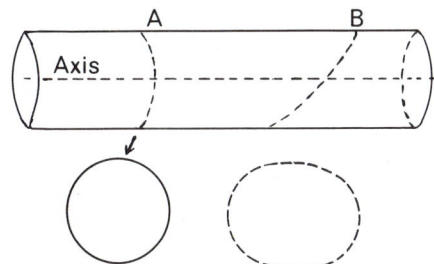

A cut made at right angles to the axis A will leave circular ends.

A cut which is not at right angles to the axis will form a curve called an **ellipse**.

## Let's Try

**1.** Which of these things are **cylindrical**?

a.

b.

c.

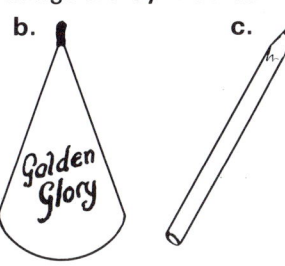

d.

e.

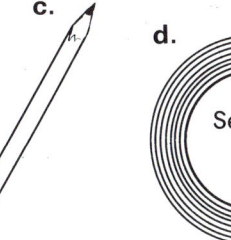

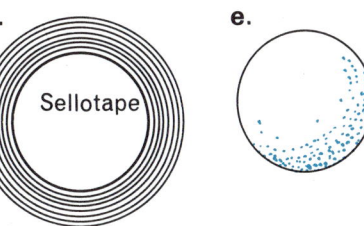

Sellotape

**2. a.** How many plane faces has a cylinder?

**b.** How many curved faces has a cylinder?

**c.** Count the number of edges on a cylinder?

**d.** Has a cylinder any vertices?

**3.** Draw the edge of the cut made on each of these cylinders.

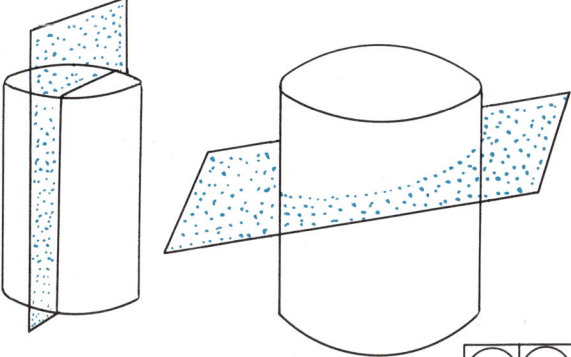

**4.** Here is a box into which 10 tins of soup will just fit. The box is 50 cm long.

**a.** Work out the perimeter of the top of the box.

**b.** Find the circumference of the top of one of the tins.

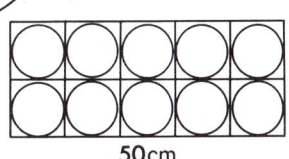

50cm

**How to Use Pie Charts**

Here is a graph which looks like a pie.
This type of graph is called a **pie chart**.
Each of the three regions is called a
**sector** of the circle.

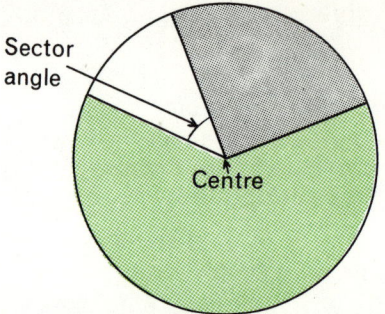

Sector
angle

Centre

**1. a.** What colour is the sector which is
$\frac{1}{4}$ of the circle?

  **b.** What colour is the sector which is
$\frac{5}{8}$ of the circle?

● In order to be able to draw pie charts you need to be able to calculate sector angles.

For example, to shade $\frac{1}{8}$ of a circle
the angle will be:

$\frac{1}{8}$ of 360° = 45°

**2.** Check that the white sector has an
angle of 45°. Is the white sector $\frac{1}{8}$ of
the full circle?

**3.** In a class of 30 pupils, 8 are
left-handed.

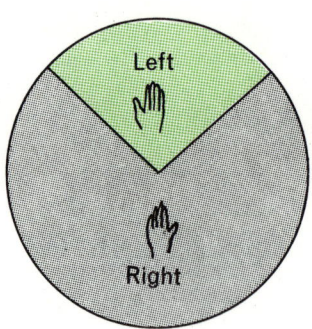

Left

Right

  **a.** How many are right-handed?

  **b.** Complete this working.

Fraction of left-handed pupils = $\frac{8}{30}$
so the sector angle = $\frac{8}{30}$ of 360°
= . . .

  **c.** Check the answer to your
calculation in part **b** by measuring
the sector angle on this pie chart.

  **d.** Work out the sector angle for
right-handed pupils.

Fraction of right-handed pupils = $\frac{22}{30}$ so sector angle = . . .

Check your answer on the pie chart.

Sometimes we need to obtain information from a pie chart. At the top of the next
page there is a pie chart which shows how Tim spends his day.

**4. a.** Check that the sector angle for "meals" is 30°.

  **b.** Complete this working.
30° as a fraction of 360° = . . . = $\frac{1}{12}$

  **c.** Complete this working to find the
number of hours spent in
"eating".

$\frac{1}{12}$ of 24 hours = . . . hours.

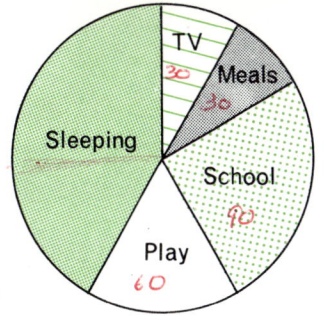

**d.** Is the sector angle for "play" twice the sector angle for "meals"? How many hours are spent in "play"?

1. Use your protractor and make calculations to help you complete this table.

**The way Tim spends his day**

| Sleeping | Play | Eating | School | TV | |
|---|---|---|---|---|---|
| | 60° | 30° | | | Angle |
| | $\frac{1}{6}$ | $\frac{1}{12}$ | | | Fraction |
| | 4 | 2 | | | Hours |

Make sure that:

**a.** the angles add up to 360°

**b.** the fractions add up to 1

**c.** the hours add up to 24

2. In a car park there are 45 cars, 20 lorries, 10 vans, and 15 scooters.

**a.** Copy and complete the following table:

| | Cars | Lorries | Vans | Scooters | **Total** |
|---|---|---|---|---|---|
| Number | | | | | |
| Fraction | | | | | |
| Sector angle | | | | | |

**b.** Draw a pie chart showing **Vehicles in the car park**.

3. In my pocket I have £1 in coins.
The way in which the pound is made up is shown on this pie chart.
Copy and complete this table:

| Type of coin | 10p | 5p | 2p | 1p |
|---|---|---|---|---|
| Sector angle | | | 108° | |
| Fraction of circle | | | | $\frac{1}{4}$ |
| Amount for each sector | | 15 | | |
| Number of coins | 3 | | | |

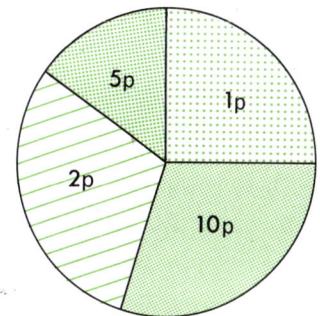

4. In a school of 780 children it was found that 156 children had light-coloured hair, 442 children had medium-coloured hair and the rest had dark-coloured hair. Draw a pie chart from this information.

65

**Cones**

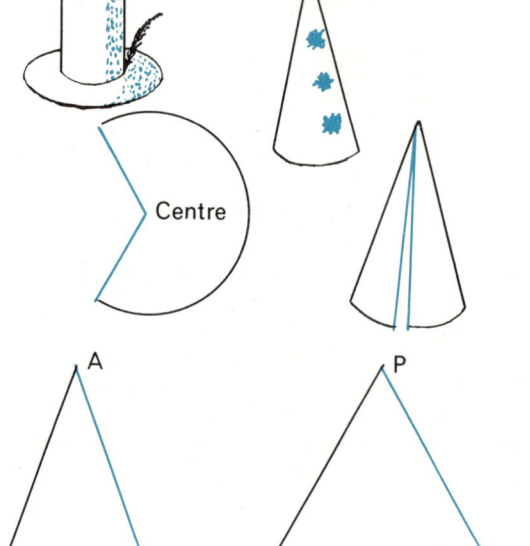

Here are two party hats which you
could make from stiff paper.

1. Which one is made from a cylinder?

   The clown's hat can be made by
   cutting out a circular piece of paper
   and:

   a. marking the centre of the circle.

   b. cutting along two radii (blue) and
      joining the two straight edges
      with tape.

2. a. Draw and cut out a circle with a
      radius of 10 cm. Make a small
      clown's hat.

   b. Cut out another circle with a
      radius of 10 cm. Make a hat which
      will fit a clown with a larger head.

   c. Measure the length of the line
      segments $\overline{AB}$ and $\overline{PQ}$.
      Is m($\overline{AB}$) = m($\overline{PQ}$)?

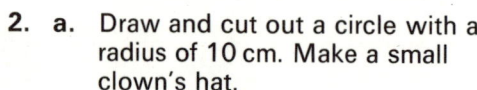

Size **7**          Size **12**

- The clown's hat is a mathematical solid which is called a **cone**.
- A line segment such as $\overline{AB}$ or $\overline{PQ}$ is called the **slant height** of the cone.

   The slant height of each of your cones is 10 cm.

3. To make a cone with a slant height of 20 cm we need to start with a circle of
   radius . . . centimetres.

4. Look carefully at the base of each of your cones. Is each base circular in shape?

- The base of a cone is a **circular** region.

   In this diagram:
   line segment $\overline{CB}$ is the **radius of the base**,
   line segment $\overline{AC}$ is the **height** of the cone.

5. Make a cut parallel to the base of one
   of your cones.
   You now have a cone and a small
   "lamp-shade".

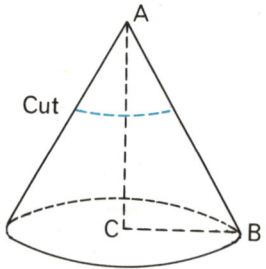

1. Copy and complete this table for the solids shown here.

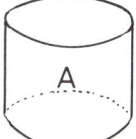 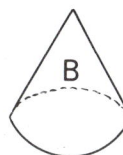 

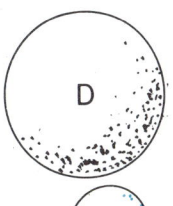

| Solid | Number of: | | | |
|---|---|---|---|---|
| | curved faces | plane faces | edges | vertices |
| A | | | | |
| B | | | | |
| C | | | | |
| D | | | | |

2. How many cones can you find in these pictures?

   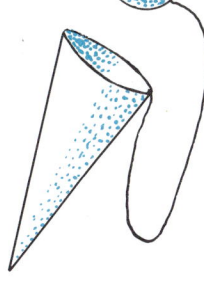

3. Use a pencil, a piece of string, and a drawing pin to draw a circle with radius 15 cm. Cut out the circle.

   a. Cut out $\frac{1}{4}$ of the circle and make a hat.

   b. Cut away $\frac{1}{2}$ of a second circle and make a hat.

   c. When you have made your hat, what does the radius of the circle become?

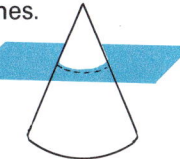

4. Draw the edge of the cut made on each of these cones.

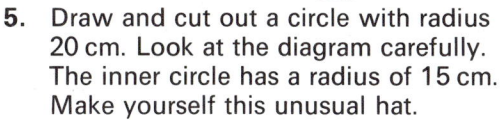

5. Draw and cut out a circle with radius 20 cm. Look at the diagram carefully. The inner circle has a radius of 15 cm. Make yourself this unusual hat.

   How many curved faces, edges, and vertices has your hat?

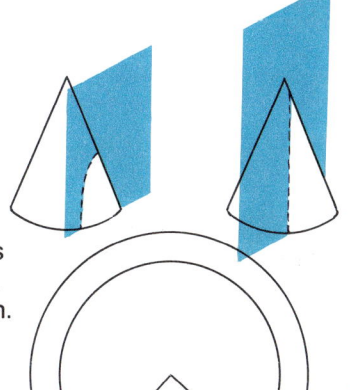

6. What kind of region is covered by a cone which is rolling on a table-top?

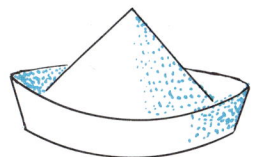

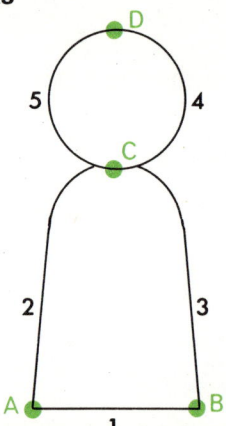

This network is a closed curve.
The network has 4 **vertices**: A, B, C, and D, and 5 **edges** numbered: 1, 2, 3, 4, and 5.

**1.** **a.** Into how many regions does the network divide the plane?

   **b.** In the following relationship, V is the number of vertices, R is the number of regions, and E is the number of edges. Does $V+R-E = 2$ for this network?

● For any network $V+R-E = 2$.

**2.** Copy the network and join vertices D and C. Complete the following sentence. "My new network has . . . vertices, 6 edges, and 4 regions and so . . . $+4-6 = $ . . ."

A map of the London Underground Railway system is a network which has many edges (lines), and vertices (stations).
It is easy to see that a train could not visit all stations without travelling along some line twice.

Here are some simpler networks.
On network (1),
3 edges meet at A, 3 edges meet at B, and 2 edges meet at C.

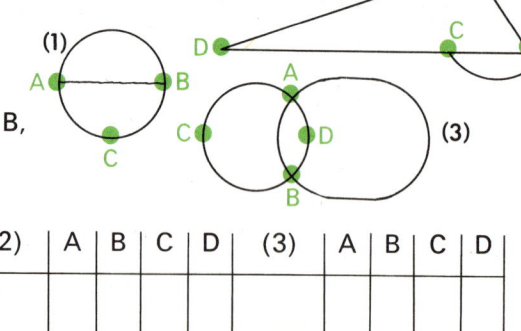

**3.** Copy and complete these tables.

| Network   (1) | A | B | C | (2) | A | B | C | D | (3) | A | B | C | D |
|---|---|---|---|---|---|---|---|---|---|---|---|---|---|
| Number of edges meeting at | 3 | 3 | 2 | | | | | | | | | | |

● Since vertex A on (1) has 3 edges meeting at it, A is called an **odd** vertex.

● Vertex C on (1) has an even number of edges meeting at it so C is called an **even** vertex.

It is possible to travel round each of the three networks without passing along any edge more than once.
I put my pencil on B and traced the route → on (1)

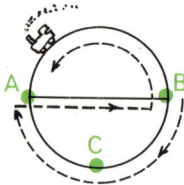

**4. a.** Find a route which travels round network (2) without travelling on any edge more than once.

This is a network on which I can't find a "**once** on any edge **only**" route. Can you find one? Try drawing the net without taking your pencil from the paper.

**5. a.** How many odd vertices has this network?

**b.** How many even vertices has this network?

It is **impossible** to find a "**once only**" route on this network.

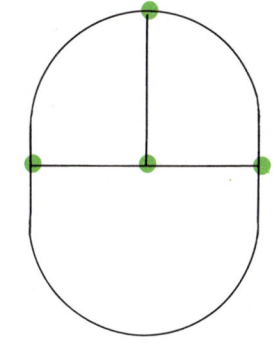

● On any network a "**once** on any edge **only**" is possible only if the number of **odd vertices** is 0 or 2.

| Let's Try |
|---|

**1.** Complete this table for the five networks. In the last column put "Yes" if you can travel round the network without passing along any edge more than once.

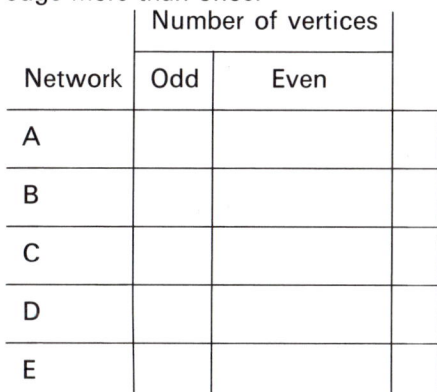

| Network | Number of vertices | | |
|---|---|---|---|
| | Odd | Even | |
| A | | | |
| B | | | |
| C | | | |
| D | | | |
| E | | | |

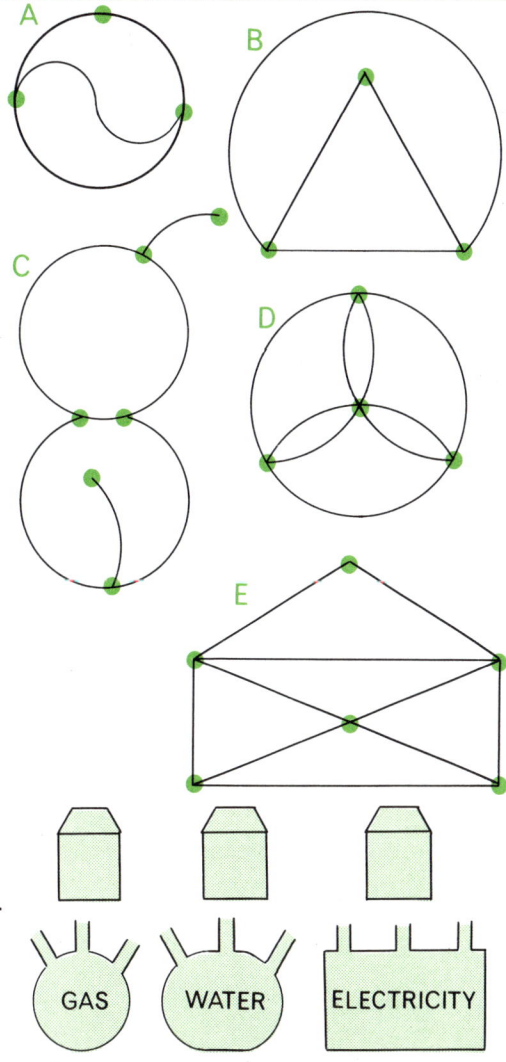

**2.** Draw some networks of your own and:

**a.** test the relationship between V, R, and E.

**b.** check whether they can be travelled round without using any edge more than once.

**3.** Here is a problem for you to investigate.

Is it possible to connect each house to the gas, water and electricity supplies without having any pipes or wires crossing?

**Multiplication of Decimals**

1. Here are two methods of multiplying $\boxed{15 \text{ by } 3\cdot5}$
   Check each method.

   **a.**
   $$1 \quad \text{of } 15 = 15$$
   $$2 \quad \text{of } 15 = 30 \quad \left.\right\} \text{Add}$$
   $$0\cdot5 \text{ or } \tfrac{1}{2} \text{ of } 15 = \ \ 7\cdot5$$
   $$\text{so } 3\cdot5 \text{ of } 15 = \overline{52\cdot5}$$
   $$15\times3\cdot5 \qquad = 52\cdot5$$

   **b.** $15\times3\cdot5 = 3\cdot5\times15$
   $$= 3\cdot5\times(10+5)$$
   $$= (3\cdot5\times10)+(3\cdot5\times5)$$
   $$= 35+17\cdot5$$
   $$= 52\cdot5$$

2. Complete the following multiplications:

   **a.** $15\times3 = \ldots$ **b.** $15\times4 = \ldots$ **c.**
   $$\begin{array}{r} 15 \\ \times35 \\ \hline \ldots \end{array}$$

   The **product** of 15 and $3\cdot5$ is greater than $15\times3$ ( $= 45$)
   and less than $15\times4$ ( $= 60$)
   $15\times35 = 525$ so, $15\times3\cdot5 = 52\cdot5$

   **d.** Is $45 < 52\cdot5 < 60$?

● When decimal points appear then
   **a.** ignore the points   **b.** complete the multiplication
   **c.** place the point in the appropriate position.

   Examples:

   **(1)** $\boxed{0\cdot3\times10}$   $3\times10 = 30$
   so $0\cdot3\times10 = 3\cdot0$ or 3

   **Check** $0\cdot3\times10 = \tfrac{3}{10}\times\overset{1}{\cancel{10}} = 3$

   **(2)** $\boxed{7\times0\cdot25}$   $7\times25 = 175$
   so $7\times0\cdot25 = 1\cdot75$

   **Check** $7\times0\cdot25 = 7\times\tfrac{1}{4} = \tfrac{7}{4} = 1\tfrac{3}{4} = 1\cdot75$

   **(3)** $\boxed{0\cdot5\times0\cdot2}$   $5\times2 = 10$
   so $0\cdot5\times0\cdot2 = 0\cdot10$ or $0\cdot1$

   **Check** $0\cdot5\times0\cdot2 = \tfrac{1}{2}\times\tfrac{1}{5} = \tfrac{1}{10}$ or $0\cdot1$

● After finding the product, you have as many decimal places as there are altogether in the two numbers being multiplied.

   $$0\cdot5\times0\cdot2 = 0\cdot10$$
   1 place   1 place   2 places

3. Follow through these examples which show how to multiply decimals:

   **a.** $\boxed{3\cdot7\times1\cdot2}$
   $$\begin{array}{r} 37 \\ \times12 \\ \hline 370 \\ 74 \\ \hline 444 \end{array}$$
   $3\cdot7\times1\cdot2$
   $= 4\cdot44$

   **b.** $\boxed{3\cdot5\times1\cdot02}$
   $$\begin{array}{r} 35 \\ \times102 \\ \hline 3500 \\ 70 \\ \hline 3570 \end{array}$$
   $3\cdot5\times1\cdot02$
   $= 3\cdot570$

Remember to ask yourself, **"Is my answer reasonable?"**.

4. John said, "$3\cdot7\times1\cdot2 = 44\cdot4$". Is John right or wrong?
   Where has John counted his decimal places from?
   Is it true to say that $3\cdot7\times1\cdot2$ is between $(3\cdot7\times1)$ and $(3\cdot7\times10)$?

Let's Try

1. {100, 500, 100000, 2000, 1000, 1000000}
   Use the elements of this set as replacements to make the following sentences true.
   Be sure that your answer is reasonable.

   **a.** $\triangle$ cm = 1 km
   **b.** $\triangle$p = £10
   **c.** $\triangle$ m = $\frac{1}{2}$ km
   **d.** $\triangle$ cm = 1 m
   **e.** $\triangle$ ml = 2 litres
   **f.** $\triangle$ g = 1 tonne

2. Without performing any further multiplication use the product in the box to help
   you write the missing number.

   $$\boxed{15\times12 = 180} \qquad \boxed{206\times8 = 1648}$$

   **a.** $15\times120 = \ldots$
   **b.** $20\cdot6\times\ldots = 164\cdot8$
   **c.** $15\times\ldots = 18$
   **d.** $2\cdot06\times8 = \ldots$
   **e.** $1\cdot5\times\ldots = 18$
   **f.** $2\cdot06\times80 = \ldots$
   **g.** $1\cdot5\times1\cdot2 = \ldots$
   **h.** $20\cdot6\times\ldots = 16\cdot48$

3. Complete the following multiplications:

   **a.** $4\cdot7\times20$
   **b.** $4\cdot7\times24$
   **c.** $83\cdot2\times1\cdot5$
   **d.** $2\cdot34\times1\cdot2$
   **e.** $4\cdot9\times1\cdot02$
   **f.** $6\cdot71\times2\cdot12$

4. This is a cuboid which has
   L = 6 cm, B = 2·5 cm, H = 1·3 cm.
   Work out the total length of 250 boxes:

   **a.** placed end to end
   **b.** side by side

   **c.** What is the height of 250 boxes
   stacked top to bottom?

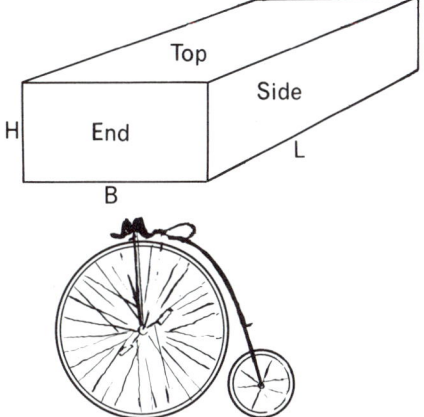

5. For one turn of the front wheel this
   bicycle travels 3·75 metres.
   How far will it travel in:

   **a.** 100 turns
   **b.** 45 turns
   **c.** 4·5 turns

/1

**Travel Graphs**

John wanted to be a cross country runner so he set off for a **4-kilometre run** one morning.

**1.** **a.** John ran for the first 2 km at a speed of 8 km per hour.
How long did this take him?

**b.** He walked the last 2 km at a speed of 4 km per hour.
How long did it take him to walk?

Here is
**A graph showing
John's progress**

Scale:
1 cm represents . . . minutes
1 cm represents . . . km

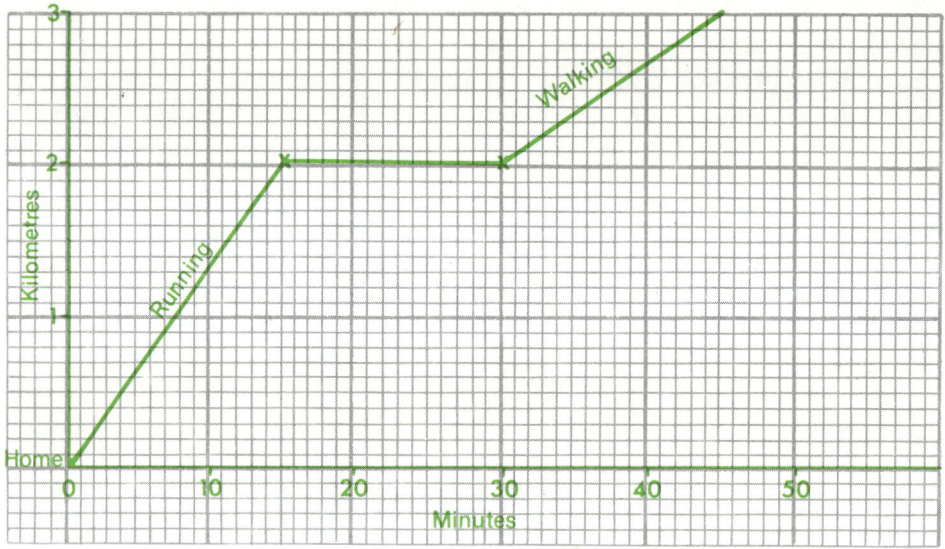

**2.** **a.** Fill in the missing numbers on the scale of the graph.

**b.** Look carefully at the graph. Is it steeper when he is walking or when he is running?

● The steeper the slope of the graph the faster the speed.

**c.** Which part of the graph shows that John rested?
For how long did he rest?

**d.** Copy and complete the graph for the 4-km journey.
How long did John's journey last?

● Graphs like the one showing John's progress are called **travel graphs**.

John's travel graph assumes that he ran steadily at 8 km per hour, and that he walked steadily at 4 km per hour.

**1.**

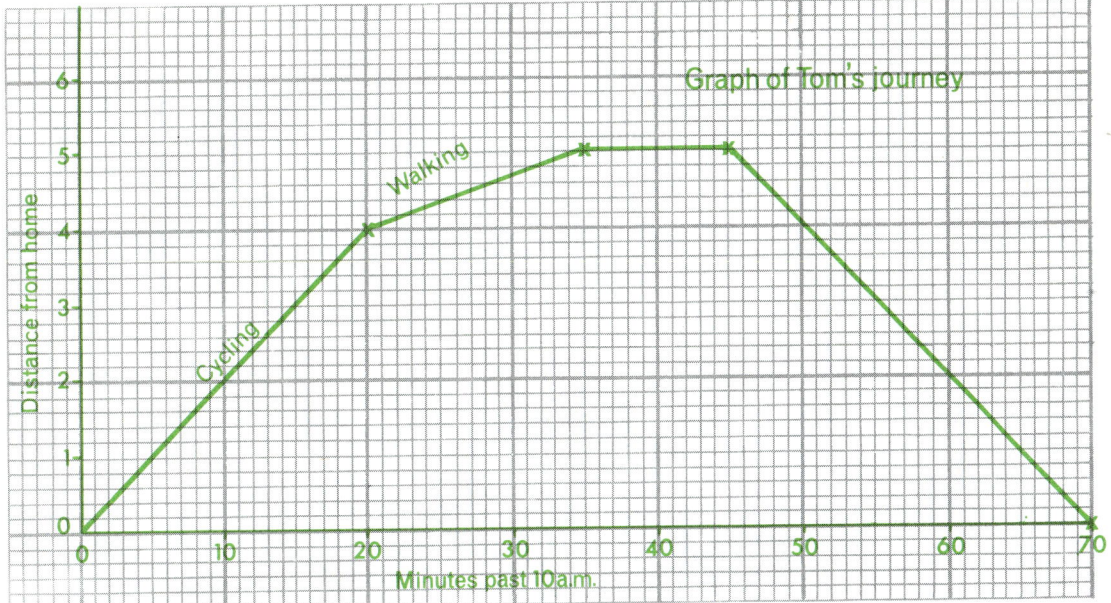

Graph of Tom's journey

Distance from home

Walking

Cycling

Minutes past 10 a.m.

**a.** How far from home is Tom at:
10.10 a.m.; 10.35 a.m.; 10.45 a.m.; 11 a.m.?

**b.** At what times is Tom the following distances from home?
3 km; 1 km; $4\frac{1}{2}$ km; 5 km

**c.** For how many minutes did he rest?

**d.** What is his speed of: cycling, and walking?

**e.** Write a story about the journey.

**2.** Draw axes similar to those in question **1**.

John sets off from home at 10 a.m. and walks quickly and steadily at 6 km per hour. Show his journey on your graph.

Tom leaves John's house at 10.20 a.m. and cycles at 12 km per hour in order to overtake John. Show Tom's journey on the graph.

At what time will Tom overtake John?

**3.** Jane decided to spend a day with her aunt who lived 16 km away. After cycling the first 8 km at 10 km per hour she rested for 10 minutes. The final stage of the journey was covered in 40 minutes. Draw a graph and find out how long the journey took.

**Volume**

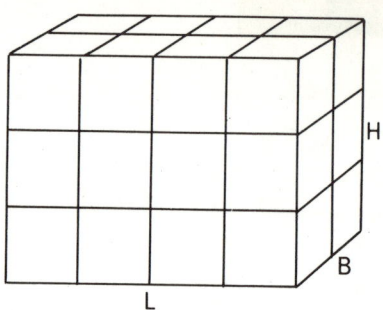

1. **a.** Arrange 24 cubes in a **cuboid** like this and then write L = ..., B = ..., H = ... "cubes".

   The volume of this **cuboid** is $4 \times 2 \times 3 = 24$ "cubes".

   **b.** Arrange your cubes in a different cuboid and check that $L \times B \times H = 24$.

● The **volume** of a cuboid is found by multiplying $L \times B \times H$.

If the lengths are in **centimetres** the volume is in **cubic centimetres**, and if the lengths are in **metres** the volume is in **cubic metres** ($m^3$).

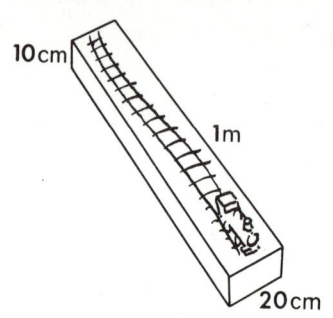

This is a box which is used to pack tracks for model railways. Its length is in metres but the other two measurements are in centimetres. Remember that L, B, and H must be in the same units, and to change metres to centimetres multiply by 100
$2 \cdot 5\,m = 250\,cm$
to change centimetres to metres divide by 100
$75\,cm = 0 \cdot 75\,m$

2. Complete this working to find the volume of the railway-track box.
   $$\text{Volume} = L \times B \times H = 1\,m \times 20\,cm \times 10\,cm$$
   $$= 100 \times 20 \times 10\,cm^3, \text{ or } 1 \times \ldots \times 0 \cdot 1\,m^3$$
   $$= \ldots cm^3 \qquad\qquad = 0 \cdot 02\,m^3$$

This box is a 1-metre cube. Try to find a fuel bunker or some other box which is approximately 1 metre by 1 metre by 1 metre.

3. Write the missing numbers.
   The fuel bunker has a volume of
   $$1\,m \times 1\,m \times 1\,m = \ldots m^3$$
   or $100\,cm \times 100\,cm \times 100\,cm = \ldots cm^3$
   $$= 10^?\,cm^3$$

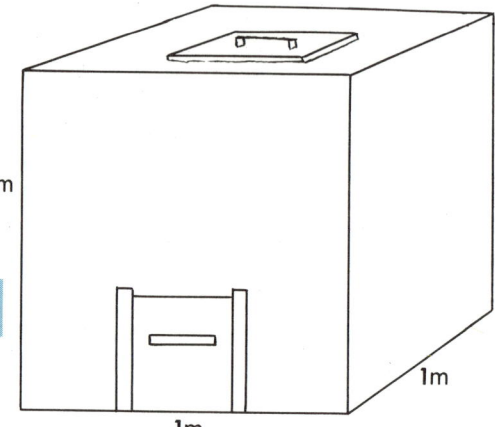

● 1000000, or $10^6$ $cm^3$ are equal to 1 $m^3$.

4. Find a 1-centimetre cube and hold it next to your fuel bunker. You would need about 1 million cubes to fill the bunker.

5. Look back at question **2** and check that
   $0 \cdot 02\,m^3 = 0 \cdot 02 \times 1000000\,cm^3$

1. Find the volume of these cuboids in cubic centimetres:

   a.                          b.                          c.

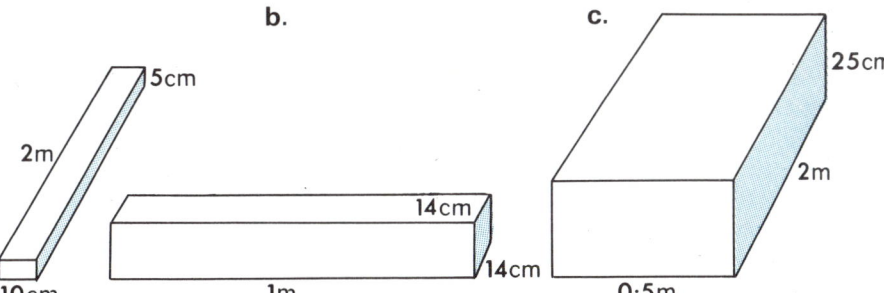

2. A match-box measures approximately 5 cm × 3·5 cm × 2 cm.

   a. What is the approximate volume of the match-box?

   b. If a match has a volume of $\frac{1}{2}$ cubic centimetre what is the greatest number of matches that could possibly go in the box?

3. This removal van has been sent to transport an estimated 35 cubic metres of household goods.

   a. Will the van hold the goods?

   b. What should be charged for the removal job if the rate is 90p per cubic metre of goods?

4. A mine shaft measuring 65 metres by 6 metres by 5 metres is to be excavated and the earth carried away in lorries which hold 11 cubic metres.

   a. What volume of earth is to be moved?

   b. How many lorry journeys will be required?

5. A swimming pool is 50 metres long, 25 metres wide, and $1\frac{1}{2}$ metres deep.

   a. How many cubic metres of water are needed to fill the pool?

   b. **1000 litres = 1 cubic metre**
      How many litres of water are contained in the pool?

6. A store-room is 5 metres long, 4·5 metres wide, and 3·5 metres high.

   a. What is the volume of the room?

   b. How many crates 1 metre by 1 metre by 1 metre can be stacked in the room? Take care with your answer.

   c. How many crates $\frac{1}{2}$ metre by $\frac{1}{2}$ metre by $\frac{1}{2}$ metre occupy 1 cubic metre? How many of these crates would just fill the store-room?

**Decimal Answers**

Here is a division of 123 by 15.

The answer can be written as:

$$8 \text{ remainder } 3 \quad \text{or} \quad 8\tfrac{3}{15}$$

8 remainder 3

$$15\overline{)123}$$
$$\phantom{15)}120$$
$$\phantom{15)12}3$$

$8\tfrac{3}{15}$ or $8\tfrac{\cancel{3}^{1}}{\cancel{15}_{5}}$ is an **exact answer**.

1. How many tenths equal 3 units?

Write the missing number. $3 = \dfrac{?}{10}$

● Instead of writing the answer to a division as a fraction we can continue the working into decimals by writing 123 as 123·0.

$$8\cdot2$$
$$15\overline{)123\cdot0}$$
$$\phantom{15)}120$$
$$\phantom{15)1}3\,0 \text{ (tenths)}$$
$$\phantom{15)1}3\,0$$

8 units × 15 = 120 units

2 tenths × 15 = 30 tenths

2. **a.** Here is another example. Follow through the working.

$$9$$
$$8\overline{)74}$$
$$\phantom{8)}72$$
$$\phantom{8)}2$$

$$9\cdot25$$
$$8\overline{)74\cdot00}$$
$$\phantom{8)}72$$
$$\phantom{8)}20$$
$$\phantom{8)}16$$
$$\phantom{8)}40$$
$$\phantom{8)}40$$

2 tenths × 8 = 16 tenths

5 hundredths × 8 = 40 hundredths

$$74 \div 8 = 9 \text{ rem. } 2$$
$$\text{or } 9\tfrac{\cancel{2}^{1}}{\cancel{8}_{4}}$$

$$74 \div 8 = 9\cdot25$$

Notice how the units are changed to tenths and the tenths are changed to hundredths.

This is an unusual example because there never seems to be an end to the division.

$$155 \div 15$$

3. **a.** Continue the division to two more decimal places.

**b.** Is $\dfrac{155}{15} = 10\tfrac{\cancel{5}^{1}}{\cancel{15}_{3}}$ ?

**c.** Is $10\cdot3 = 10\tfrac{1}{3}$?

**d.** Write 0·33 as a fraction. $\dfrac{?}{100}$ Is $10\cdot33 = 10\tfrac{1}{3}$?

$$10\cdot3?$$
$$15\overline{)155\cdot00}$$
$$\phantom{15)}15$$
$$\phantom{15)}5\,0$$
$$\phantom{15)}4\,5$$
$$\phantom{15)}50$$
$$\phantom{15)}?$$

A decimal such as 10·33333 is called a **recurring decimal** and is written 10·3̇.

When dividing in decimals it is usually sufficient to work to three decimal places at the most.

4. Check these answers by working long division.

   **a.** $123 \div 12 = 10 \cdot 25$
   or $= 10\frac{1}{4}$

   **b.** $142 \div 14 = 10 \cdot 142$ (**approximately**)
   or $= 10\frac{1}{7}$ (**exactly**)

Let's Try

1. Complete these divisions. Work to 3 decimal places at the most.

   **a.** $49 \div 4$      **b.** $49 \cdot 7 \div 4$      **c.** $127 \div 11$

   **d.** $423 \div 15$      **e.** $42 \cdot 3 \div 15$      **f.** $327 \cdot 52 \div 22$

2. A pile of similar books is $1 \cdot 25$ metres high. How thick is each book in metres if there are:

   **a.** 10 books in the pile      **b.** 15 books in the pile

   **c.** Write your answers in centimetres.

3. This diagram shows steps used for mounting the school platform. The tread equals the rise, and the length of carpet used is $1 \cdot 2$ metres.

   Find the length of rise, and of tread:

   **a.** as a decimal of a metre

   **b.** in centimetres

   **c.** Work out the height of the platform above the floor.

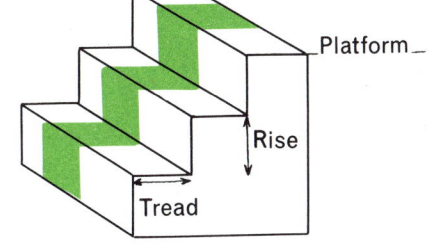

4. Andrew's train layout has a total track length of $10 \cdot 5$ metres and is made up from 42 sections of standard length. Work out the length of each section:

   **a.** as a decimal of a metre

   **b.** in centimetres

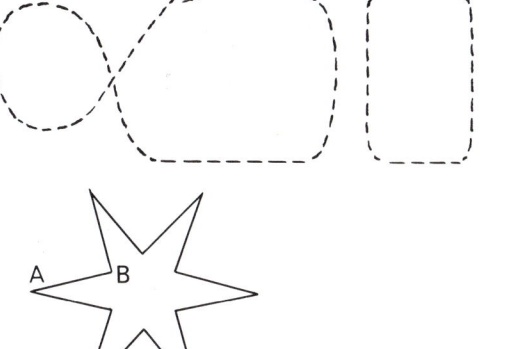

5. 135 centimetres of wire is bent to form this shape which has equal sides. Find as a decimal:

   **a.** the length of segment $\overline{AB}$.

   **b.** the length of a segment if the star has 8 points.

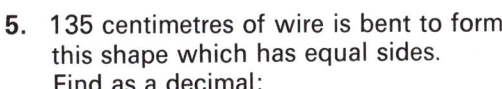

Whose hand has the larger area, yours
or your neighbour's?

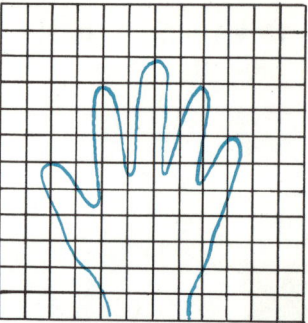

1.  **a.**  Place your hand on a piece of
        centimetre-squared paper and
        draw the shape of your hand.
    **b.**  Count the number of whole squares
        and piece together parts of squares
        to find the approximate area of
        your hand in square centimetres.
    **c.**  Now see whether your hand has a
        larger area than your neighbour's.

●  Area is a measure of the surface in a region.

The area of the region you have measured is "hand-shaped".

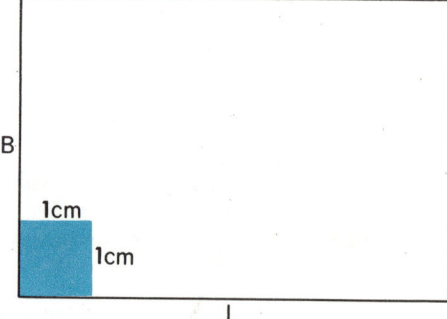

2.  Use your ruler to measure:
    **a.**  the length of this rectangle in
        centimetres
    **b.**  the breadth of this rectangle
    **c.**  Draw the rectangle and then draw
        lines which divide it into
        centimetre squares.
    **d.**  Is your region divided into
        $6 \times 4 = 24$ squares?
    The area (A) of your rectangle is
    $6\,cm \times 4\,cm = 24$ square centimetres (cm²).

●  The **area** (A) of a rectangular region is found by multiplying the length by the
    breadth.
●  $L\,cm \times B\,cm = A\,cm^2$      $L$ metres $\times B$ metres $= A$ square metres (m²).

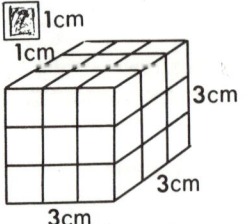

3.  How many 1-cm square stamps
    would be needed to cover:

    1 face of this cube.
    2 faces of this cube.
    3 faces of this cube.
    all 6 faces of this cube.

●  The total **surface area** of a cube is $6\times$ the area of one face.

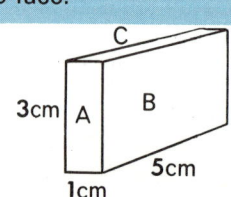

Find a cuboid like this one on which
L, B, and H are all different.
4.  Check that:
    **The total surface area**
    $= [2 \times m\ (\text{face A})] + [2 \times m\ (\text{face B})] + [2 \times m\ (\text{face C})]$

1. Work out the area of these rectangles in the units stated.

   **a.** L = 15 cm   B = 7 cm
   (square centimetres)

   **b.** L = 20 m   B = 16 m
   (square metres)

   **c.** L = 2 m   B = 15 cm
   (square centimetres)

   **d.** L = 8 m   B = 25 cm
   (square metres)

2. Write these areas in square centimetres for this cuboid:

   **a.** m(ABCD)     **b.** m(CDEF)
   **c.** m(BCFG)     **d.** m(total surface)

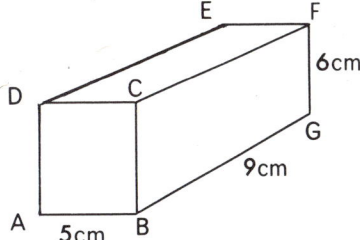

3. Work out the total surface area, and the volume of these cuboids:

   **a.**                    **b.**                    **c.**

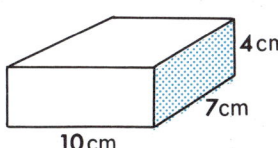
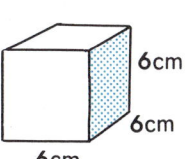
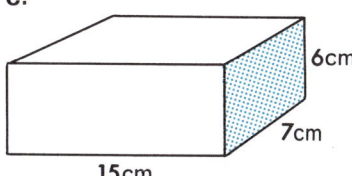

Did you notice anything special about the area and volume of **b**?

4. Here is a plan for a lounge floor. Work out:

   **a.** the area of the floor

   **b.** the area of the carpet in square metres

   **c.** the area of the carpet in square centimetres

   **d.** Complete this sentence.
   1 square metre = . . . square centimetres.

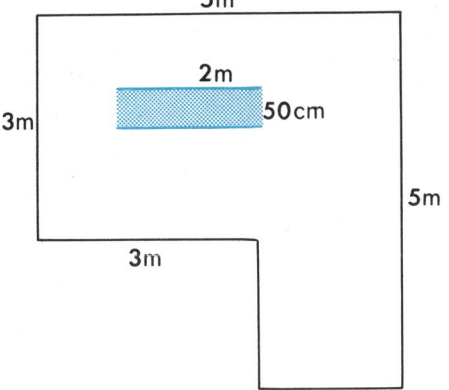

5. This is a 3-cm cube which has been built from 1-cm cubes. The **large cube** has been painted blue.

   **a.** How many square centimetres have been painted blue?

   **b.** How many 1-cm cubes have: 3 faces blue, 2 faces blue, and 1 face blue?

   **c.** How many 1-cm cubes have no faces blue?

**Curved Graphs**

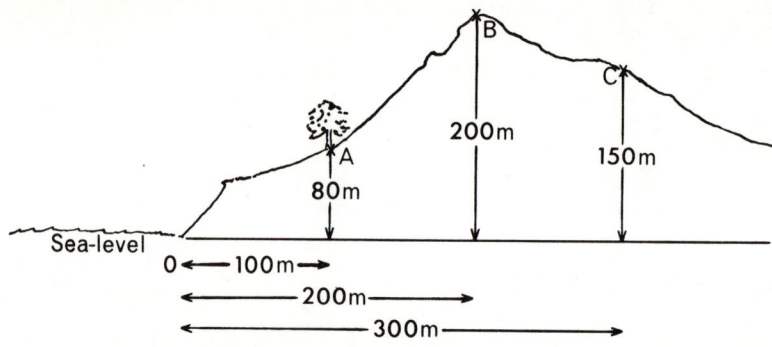

Here is a picture of a hill at the edge of the sea. Let us call the point where the sea and the hill meet the **origin, O**.

When a climber reaches the tree he has moved 100 metres **horizontally**.

1.  **a.**  How high is the climber above sea-level when he reaches the tree?

    **b.**  Complete the following sentences.
    The climber has travelled . . . metres horizontally when he is at the top of the hill.
    At the top of the hill he will have reached a height of . . . metres.

This is a graph of the points:
A (100, 80), B (200, 200),
C (300, 150)

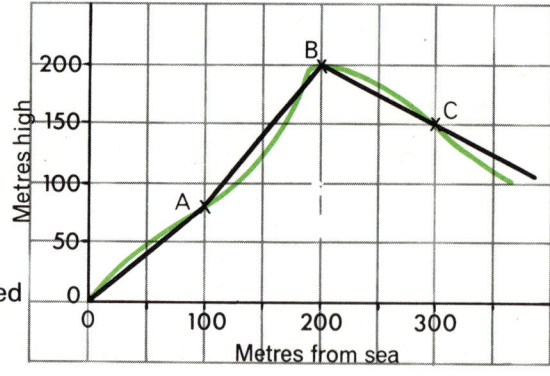

2.  Does the hill really look like the line ———?
    Will the climber walk on the line segment $\overline{AB}$?

    It would be an unusual hill if it looked like the black line and had a sharp point at the top.

●  The graph of the hill should be curved. The green outline is more correct than the black line.

3.  **a.**  Copy this table and then complete it after finding the area of each square.

| Length of side | 0 | 1 | 2 | 3 | 4 | 5 | centimetres |
|---|---|---|---|---|---|---|---|
| Area of square |  |  |  |  |  |  | square centimetres |

    **b.**  Cut out squares to match the ones in the table and watch the way the area grows as the length of side increases.

    **c.**  Complete the graph at the top of the next page. Use your table.

This is part of your graph. The black line does not give us a true picture of what happens as the side of the square is increased.

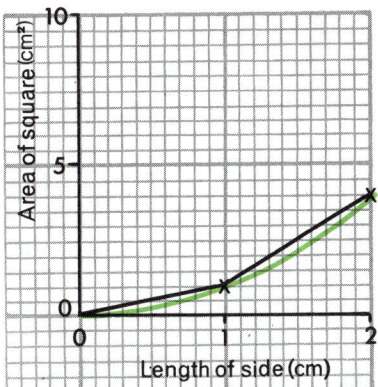

4.  **a.** The area of a $\frac{1}{2}$-cm square is $\frac{1}{4}$ cm². Check this on the graph. Does the point $(\frac{1}{2}, \frac{1}{4})$ fall on the black line or the green curve?

    **b.** The area of a $1\frac{1}{2}$-cm square is $2\frac{1}{4}$ cm². Check this on the graph. Does the point $(1\frac{1}{2}, 2\frac{1}{4})$ fall on the black line or the green curve?

- The green curve is a correct graph showing the way squares grow as the length of side increases.

5.  Now complete your own curved graph showing the **area of a square for a given length of side**.

Let's Try

1.  **a.** Using whole number values only, how many different rectangles can you make with an area of 24 square centimetres? Your rectangle could be 24 cm by 1 cm
    or   1 cm by 24 cm.

    **b.** Cut the rectangles you have found in part **a** from coloured paper and stick them on axes like the one in this diagram. Draw the smooth green curve.

    **c.** Does the point $(16, 1\frac{1}{2})$ fall on your curve? What is the area of a rectangle which is 16 cm long and $1\frac{1}{2}$ cm wide?

2.  Mary decides to make a rectangular table-mat. The perimeter of the base is to be 40 centimetres.
    $L+B = 20$ cm

    **a.** Complete this table for the length, breadth, and area (A) of the base.

| L | 1 | 2 | 4 | 6 | 8 | 10 | 12 | 14 | 16 | 18 | 19 | 20 | cm |
|---|---|---|---|---|---|----|----|----|----|----|----|----|----|
| B | 19 | 18 | | | | | | | | | | | cm |
| A | 19 | 36 | | | | | | | | | | | cm² |

    **b.** Draw a curved graph showing the **area** for a given **length** from your table, using a scale of 1 cm representing 1 cm of length, and 1 cm representing 5 cm². Which rectangle has the greatest area?

## Wholesale and Retail

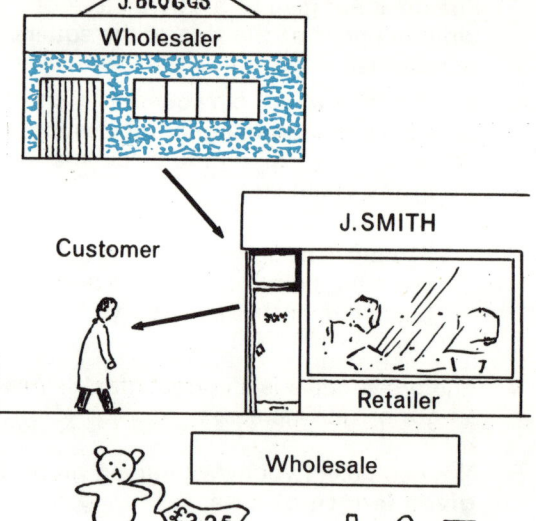

J.BLOGGS
Wholesaler

Customer

J.SMITH

Retailer

You have often been to a shop to buy something from the shopkeeper.

1. Before you can buy from a shop, the shopkeeper must buy in his goods or **stock**. Where do you think your shopkeeper buys his stock from? He might tell you.

• A shopkeeper buys his stock in large quantities from a **wholesaler**. A shopkeeper sells his stock in smaller quantities and is called a **retailer**.

2. These two items of stock are available in the wholesaler's warehouse. Write the missing numbers in the following sentences:

Wholesale

£3·25

£2·60

Retail

£3·75

£3

a. The retailer will pay
   £. . .×10 = £32·5 for ten dolls, and
   £. . .×10 = £. . . for ten trains.

   The retailer will charge £3·75 for a doll and £3 for a train.

b. The retailer is charging an extra
   . . .p on a doll, and . . .p on a train.

• The extra money the retailer charges is called his **profit**.

c. Check that the retailer makes 500p or £5 profit on 10 dolls, and 400p or £4 profit on 10 trains.

The wholesaler needs help with his arithmetic because he is selling a large number of items of stock.
To help him work out the cost he uses a *Ready Reckoner* like this one:

The symbol @ means "at".

| @ | Number of articles | | | | | | | | | | | | |
|---|---|---|---|---|---|---|---|---|---|---|---|---|---|
| @ | 1 | 2 | 3 | 4 | 5 | 6 | 7 | 8 | 9 | 10 | 20 | 30 | 40 |
| 2p | 2 | 4 | 6 | 8 | 10 | 12 | 14 | 16 | 18 | 20 | 40 | 60 | |
| 3p | 3 | 6 | 9 | 12 | 15 | 18 | 21 | | | | | | |
| 4p | | | | | | | | | 36 | 40 | 80 | | |

The grey-shaded square shows that 9 articles @ 4p cost 36p.
The blue-shaded square shows that 20 articles @ 4p cost 80p.

3. Complete this sentence which shows you how to use the *Ready Reckoner* to find the cost of 29 articles @ 4p each.

   "29 articles @ 4p each cost 80p+36p = . . .p = £. . .

**1.** **a.** Make a *Ready Reckoner* like the one on the opposite page.
Extend it to 50 articles and up to 9p.

**b.** Use your *Ready Reckoner* to help you complete these bills:

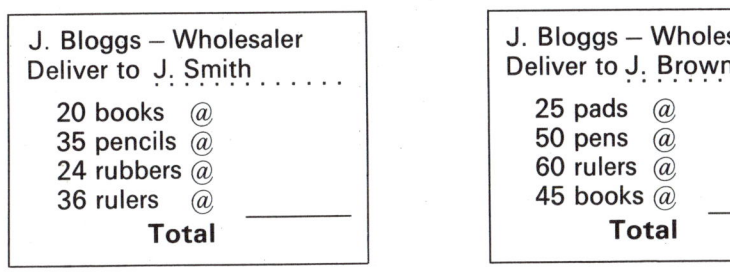

| J. Bloggs – Wholesaler | |
| --- | --- |
| Deliver to J. Smith . . . . . . | |
| 20 books @ | |
| 35 pencils @ | |
| 24 rubbers @ | |
| 36 rulers @ | _____ |
| **Total** | |

| J. Bloggs – Wholesaler | |
| --- | --- |
| Deliver to J. Brown . . . . . . | |
| 25 pads @ | |
| 50 pens @ | |
| 60 rulers @ | |
| 45 books @ | _____ |
| **Total** | |

| | | | | | |
| --- | --- | --- | --- | --- | --- |
| **Wholesale** 8p | 3p | 2p | 4p | 6p | 7p |

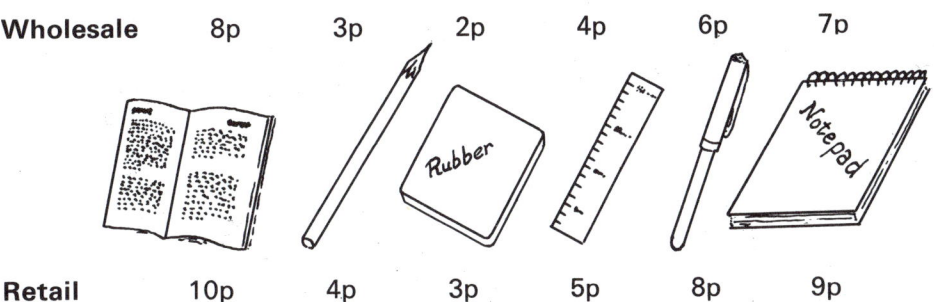

| | | | | | |
| --- | --- | --- | --- | --- | --- |
| **Retail** 10p | 4p | 3p | 5p | 8p | 9p |

**c.** Work out the profit for J. Smith and J. Brown if they sell all the stock they
have bought.

**2.** **A garage pays**

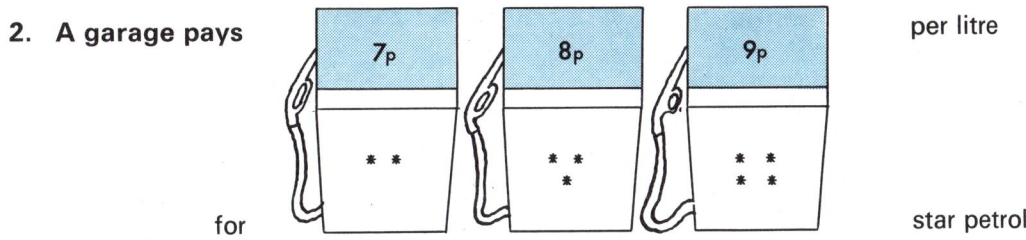

per litre

for ...... star petrol

**a.** How much will the garage owner pay for delivery of:
1000 litres of 2 star petrol, 2500 litres of 3 star petrol, and
2000 litres of 4 star petrol?

**b.** If he makes a profit of 1p per litre, how much profit does he make on the
delivery?

**3.** Farmer Giles sells his milk to a retailer at 5p per litre.
The retailer sells the milk to housewives at 8p per litre.

**a.** How much profit does the retailer make on 800 litres of milk?

**b.** Draw a Ready Reckoner graph to show:
how much Farmer Giles receives for his milk, and how much the retailer
receives for the milk.

**Time**

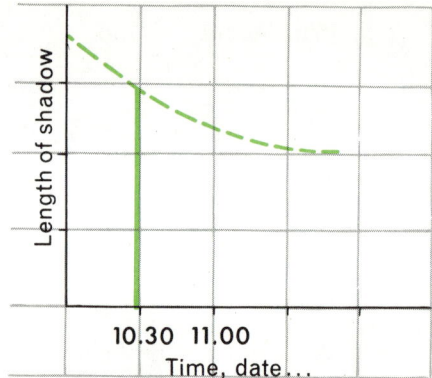

1.  a.  Use a ball of plasticine to fix a
        ruler or strip of wood in a vertical
        position on a sheet of card.
        Place your card on the ground in
        an open space.
        **Choose a sunny day.**

    b.  Complete this table:

| Time of day | 10.30 | 11.00 | 11.30 | 12.00 | 12.30 | 13.00 | 13.30 | 14.00 |
|---|---|---|---|---|---|---|---|---|
| Length of shadow in centimetres | | | | | | | | |

    c.  Draw a vertical line graph
        showing:

        **The length of shadow
        at various times**

    d.  Join the tops of your lines with
        a curved line.

    e.  Will the shadow be very long or
        very short at the times 8.00 and
        20.00 if the sun is shining?

● **The shadow will be shortest when the
sun is at its highest point and longest
when the sun is rising or setting.**

2.  a.  Cut a piece of string about
        1 metre long. Attach a small bag
        to the end of the string and hang
        it from a nail.

    b.  Put a marble in the bag and use
        a watch to time 20 swings of
        your pendulum.

    c.  Complete this table:

| Number of marbles in bag | 1 | 2 | 3 | 4 | 5 |
|---|---|---|---|---|---|
| Time for 20 swings | | | | | |

Bob

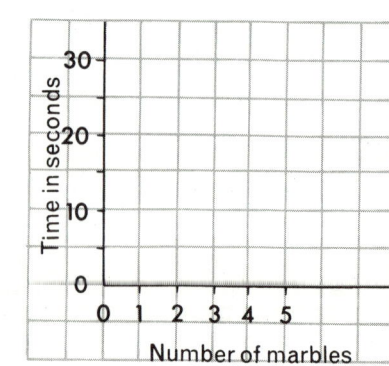

    d.  Draw a graph from your table.

    e.  Would the time for 20 swings change if you put 10 marbles in the bag?

    You can add as many marbles as you like, provided that the string does not break,
    without changing the time for 20 swings.

● **The time of swing of a pendulum is not altered by the size of weight of the bob.**

On the previous page your **units of time** were **hour, minute,** and **second.**
When working with time take care to select a suitable unit.
For example:
Jane wished to graph the weight of her guinea pig as it grew. She would not
weigh it every **minute** or even every **day.** She would weigh it every **week** or every
**month.**

● **Second, minute, hour, day, week, month,** and **year** are all units of **time.**

Let's Try

1. Which would be the most suitable unit of time for a journey from Liverpool to
New York using:

a.    b.    c.

2. Use your library to find the meaning of:

   a. decade    **b.** perennial    **c.** biennial    **d.** British Summer Time

   **e.** Greenwich Mean Time    **f.** lunar month    **g.** calendar month

3. **a.** Make a pendulum 120 centimetres long and time 100 swings.
   Complete this working:

   Time for **one** swing $= \dfrac{?}{100} = \dots$ seconds.

   **b.** Alter the length of your pendulum in order to complete this table and then draw
   a graph from your results.

   | Length of pendulum (cm) | 120 | 110 | 100 | 90 | 80 | 70 |
   |---|---|---|---|---|---|---|
   | Time for 1 swing | | | | | | |

   **c.** Use your graph to find the length
   of pendulum which has a
   "1-second swing".

4. **a.** The sun rises in the East. In which
   direction does the sun set?

   **b.** In which direction was the
   shortest shadow of your stick,
   used for question **1** on the
   opposite page, pointing?

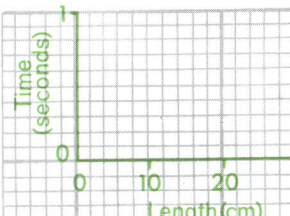

**Division by Decimals**

| 0 | 0 | 8 | 7 | · | 9 | km |

I was able to drive my car 87·9 kilometres on 10 litres of petrol.

**1. a.** John performed this division to find out how far my car travelled on 1 litre of fuel.
Can you tell John a quicker method of dividing by 10?

**b.** Divide 87·9 by 10.

```
      8·79
10)87·90
    80
    ──
     7 9
     7 0
     ──
       90
       90
       ──
```

Sometimes you will need to use long division when dividing a decimal. Look at John's working again. He wrote 87·90 for 87·9.

**2.** Is 87·900000 equal to 87·9?

A further test on my car showed that I drove 32·5 km on 2·5 litres of petrol, so on 1 litre I drove $\frac{32·5}{2·5}$ km.

**3.** Check and complete this working by cancelling.
$$\frac{3·25}{2·5} = \frac{3·25}{2·5} \times \frac{10}{10} = \frac{3·25 \times 10}{2·5 \times 10} = \frac{32·5}{?} = ?$$
The same division can be set out like this:
3·25÷2·5 = 32·5÷25 = . . . and completed by long division.

● When dividing by a decimal multiply by 10, 100, 1000, . . . to change it to a whole number. Remember that the number being divided must be multiplied by the same multiple of 10.

For example:

**4. a.** 3·12÷2·5

2·5×10 − 25
so multiply both numbers by 10.
3·12÷2·5
= 31·2÷25

```
     1· ?
25)31·200
   25
   ──
    6 2
    ?
```

**b.** 3·7÷1·05

1·05×100 = 105
so multiply both numbers by 100.
3·7÷1·05
= 370÷105

```
      3·5 ?
105)370·000
    315
    ───
     55 0
     52 5
     ────
       ?
```

Complete the working on these examples and finish off by writing:
3·12÷2·5 = 1·248     3·7÷1·05 = 3·523

**c.** Which of the two answers: **a** or **b**, is exact?

● To show when an answer is **not** exact write, "to △ decimal places".

For example:

(1) 10÷3 = 3·333
    (to 3 decimal places)

(2) 10÷3 = 3·33
    (to 2 decimal places)

Let's Try

1. Complete these divisions to at the most 3 decimal places.
   Show which answers are not exact.

   **a.** 10÷2·5        **b.** 10·5÷2·5        **c.** 102÷0·5
   **d.** 13·25÷0·8     **e.** 2·45÷1·4        **f.** 100÷1·5

2. Mother's knitting pattern says,
   "3 stitches to a cm and 4 rows to a cm". Rows

   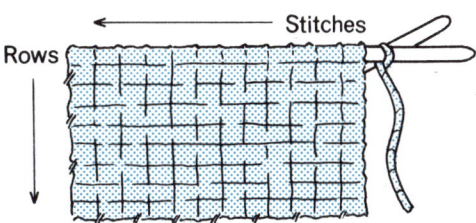

   Stitches

   **a.** After putting 26 stitches on her
   needle they measure 8 cm. How
   many stitches is this to the
   centimetre?

   **b.** By the time mother's knitting was
   7·4 cm long she had knitted 30
   rows. How many rows is this per
   centimetre?

3. John's clockwork train travels 2·4 times round a 4-metre track before coming to
   rest.
   If the train is travelling for 3 minutes:

   **a.** how long does it take for each circuit?

   **b.** how far does the train travel?

4. A paper clip is made from 8·5
   centimetres of wire.

   How many paper clips can be made
   from the following lengths of wire?

   **a.** 85 cm        **b.** 50 cm
   **c.** 1 metre      **d.** 2 metres

5. A long sausage is twisted every
   14·5 cm to make smaller sausages.

   How many sausages can be made
   from a long sausage which is:

   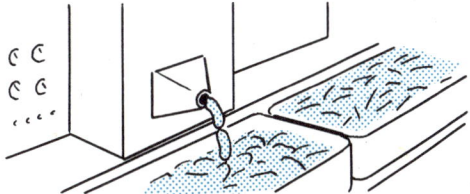

   **a.** 290 cm long        **b.** 10 metres long

1. Each of these sentences has a missing
   number. Write each number.

   1 child has . . . hands.
   . . . children have 4 hands.
   3 children have . . . hands.
   . . . children have 20 hands.

2. Is it true that the number of hands
   equals 2 times the number of
   children?
   The **ratio** of children to hands is
   1 to 2, **or** $\frac{1}{2}$ and the **ratio** of hands to
   children is 2 to 1, **or** $\frac{2}{1}$.

3. Mr. and Mrs. Jones have 2 girls and
   3 boys. Complete these sentences.

   The ratio of girls to boys is 2 to . . ., or $\frac{?}{?}$.

   The ratio of boys to girls is 3 to . . ., or $\frac{?}{?}$.

   Here are two sticks which are
   casting shadows.

4. Complete these sentences.
   a. The ratio of the length of stick A
      to the length of its shadow
      is 1 to . . .
   b. The ratio of the length of stick B
      to the length of its shadow
      is $1\frac{1}{2}$ to . . ., or $\dfrac{1\frac{1}{2}}{?} = \dfrac{1}{2}$

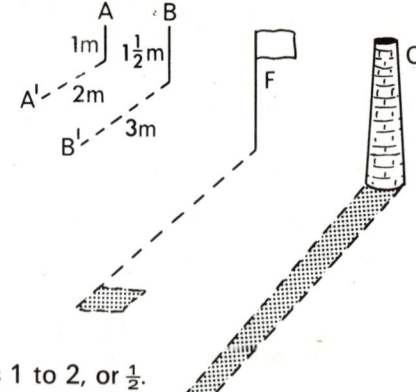

   At this particular time of the day the
   length of the shadow is twice the
   height of the object casting it.
   The ratio of **object** height to shadow length is 1 to 2, or $\frac{1}{2}$.

5. The **length of the shadow** of the flag-pole (F) is 25 metres so the height of the
   flag-pole is $25 \times \frac{1}{2} = $ . . . metres.

   The ratio of shadow length to **object** height is 2 to 1, or $\frac{2}{1}$.

6. The height of the chimney (C) is 50 metres so the length of its shadow will be
   $50 \times \frac{2}{1} = $ . . . metres.

   In questions **5** and **6** you multiplied by the **ratio**. Take care to use the correct fraction
   when working out problems with ratios.

**7. a.** Is the shadow cast by an object always twice as long as the object is high?

**b.** Complete this ratio.
The ratio of the height of the tree to the length of the shadow is
$$\frac{AB}{BC} = \frac{?}{?}$$

**1.** Write the ratio of the length of the grey rod to the length of the green rod:

**a.**                    **b.**                    **c.**

**2.** Write the ratio of the length of the green rod to the length of the grey rod in question **1**.

**3.** Write the ratio of set A to set B.

**a.**                                    **b.**

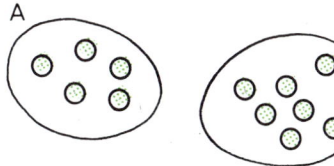                     

**4.** At a time when a gate-post casts a shadow of length 1·5 metres the shed casts a shadow which is 10·5 metres long.

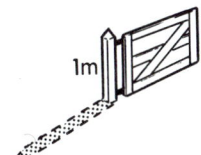      

**a.** What is the ratio of the height of the gate-post to its shadow length?

**b.** Work out the height of the shed in metres.

**c.** If the flag-pole is 9 metres high, how long will its shadow be?

**d.** Complete these ratios:
$$\frac{\text{Height of gate-post}}{\text{Height of flag-pole}} = \frac{?}{?} \qquad \frac{\text{Length of gate-post shadow}}{\text{Length of flag-pole shadow}} = \frac{?}{?}$$

89

**Congruence and Similarity**

1.  Which of these animals

    **a.**  look alike?

    **b.**  look alike and are the same size?

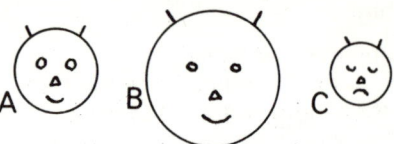

---

● Objects which are the same shape and the same size are **congruent**.

---

Faces A and D are congruent.

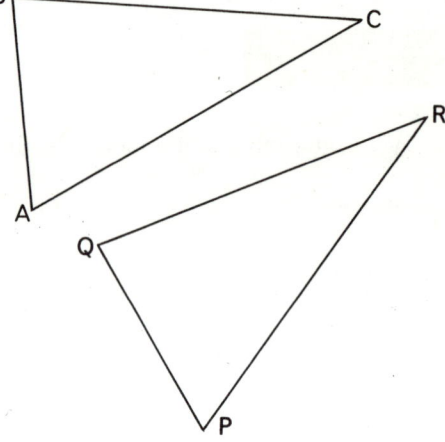

2.  **a.**  Use a protractor to help you
    write the missing angles.

    m($\angle$ BAC) = . . .    m($\angle$ QPR) = . . .

    m($\angle$ ABC) = . . .    m($\angle$ PQR) = . . .

    m($\angle$ BCA) = . . .    m($\angle$ QRP) = . . .

    **b.**  Write the missing lengths.

    m($\overline{AB}$) = . . .    m($\overline{PQ}$) = . . .

    m($\overline{BC}$) = . . .    m($\overline{QR}$) = . . .

    m($\overline{PR}$) = . . .    m($\overline{AC}$) = . . .

Triangle ABC is the same shape and
size as triangle PQR.
The two triangles are **congruent**.

Look carefully at these two photographs.

3.  Are the pictures:

    **a.**  the same to look at?

    **b.**  taken from the same negative?

    **c.**  **congruent**?

    The two pictures are described as
    **similar**.

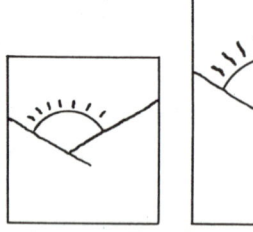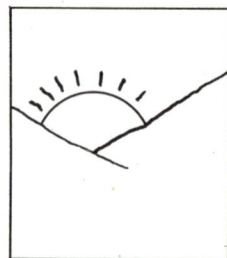

---

● Regions which are the same shape but not necessarily the same size are called
**similar**.

---

4.  Trace these triangles and cut them out. By matching angles find out which two
    triangles are **similar**.

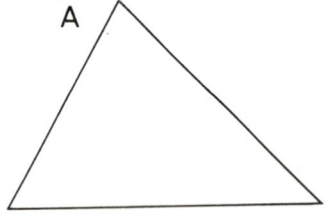  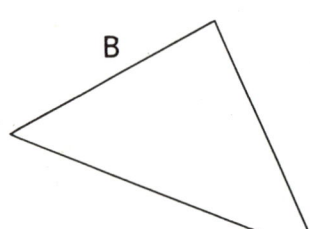  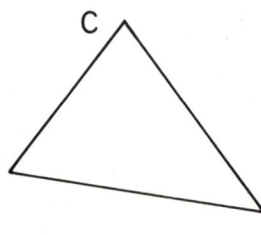

**5. a.** Write the missing angles:

∠ (1) = ...  ∠ (11) = ...
∠ (2) = ...  ∠ (12) = ...
∠ (3) = ...  ∠ (13) = ...
∠ (4) = ...  ∠ (14) = ...

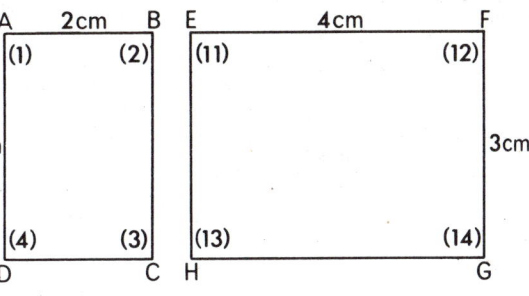

The rectangles **ABCD** and **EFGH** are not congruent since they are not the same size.
The rectangles are not similar since one is not an enlargement of the other.

Let's Try

**1. a.** Use a ruler and protractor to help you find two pairs of similar figures.

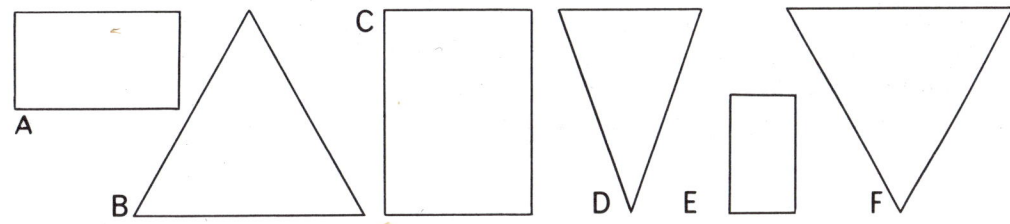

**b.** Which pair of similar figures are also congruent?

**2.** Copy this picture so that you have:

**a.** a shape which is congruent to this one, and

**b.** a shape which is twice as large and similar to this one.

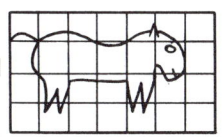

**3.** A **pantograph** is an instrument which is used to make enlargements.
Make a pantograph from loosely jointed Meccano strips and use it to draw similar figures.

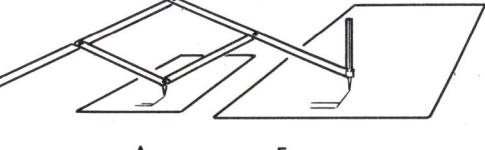

**4.** Here is a regular hexagon. Write:

**a.** the number of triangles similar to AGF.

**b.** the number of triangles congruent to AGF.

**c.** the shapes which are congruent to region AGCB.

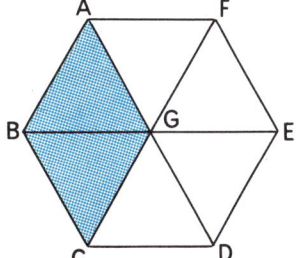

**Area of Triangular Regions**

Rectangle ABCD has:

**length** = 4 cm    **breadth** = 3 cm
**area** = 4 cm×3 cm = 12 cm²

**1.** **a.** Copy the rectangle and draw triangle PDC.

Cut up your rectangle into the four small triangles and check that:

**b.** region (1) fits exactly over region (2).

**c.** region (3) fits exactly over region (4).

**d.** area [region (1)+region (4)] = area [region (2)+region (3)]

The area of the rectangle = m($\overline{DC}$) × m($\overline{BC}$)
The area of the triangle = $\frac{1}{2}$ [m($\overline{DC}$) × m($\overline{BC}$)]

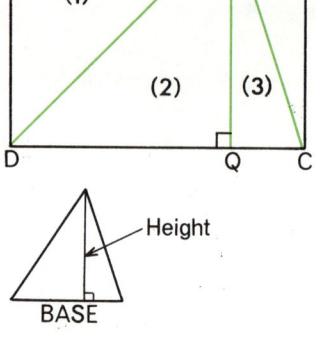

- m($\overline{DC}$) is called the **base** of the triangle, and m($\overline{BC}$) is called the **height** of the triangle.
- The area of a triangle = $\frac{1}{2}$ × **base** × **height**

**2.** **a.** Write the missing numbers.
Base of triangle ABC is $\overline{BC}$.
m($\overline{BC}$) = . . . cm
Height of triangle ABC is $\overline{AQ}$.
m($\overline{AQ}$) = . . . cm

**b.** Copy the triangle onto squared paper and cut it into the four regions.

**c.** Rearrange the four pieces to form a rectangle. What is the area of the rectangle?

**d.** Is the area of your rectangle equal to the area of triangle ABC?

The area of triangle ABC = $\frac{1}{2}$ × $\overline{CB}$ × $\overline{AQ}$ = $\frac{1}{2}$ × 3 cm × 4 cm = 6 cm².

By splitting polygons into triangular regions you can find the area of the polygons.

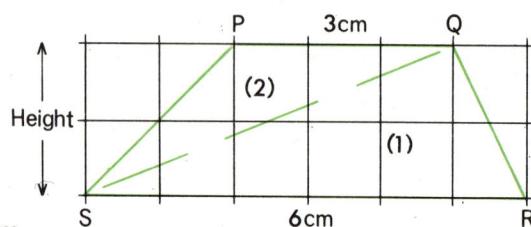

**3.** Complete these sentences.

  **a.** The height of triangle (1) = . . . cm.
  **b.** The base of triangle (1) = . . . cm.

**c.** Triangle (2) has a height of . . . cm and a base of 3 cm.

**d.** The **area** of triangle (1) $= \frac{1}{2} \times \ldots \times 2\,cm = \ldots cm^2$.

**e.** The **area** of triangle (2) $= \frac{1}{2} \times 3\,cm \times \ldots = \ldots cm^2$.

**f.** Check that the total area of trapezium PQRS is 9 cm².

Let's Try

**1.** Find the area of each of the following regions.
The lengths shown are all in centimetres.

**a.**

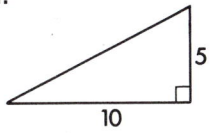

**b.**

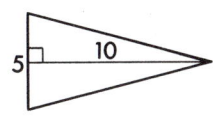

**c.**

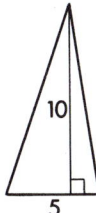

**d.**

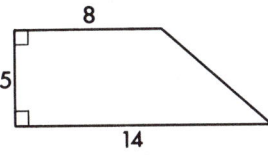

**e.**

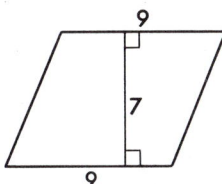

**f.**

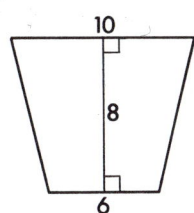

**2.** This is a picture of my garden shed which is 3 metres high. The roof of the shed is tiled and so I have to paint the two walls and the two ends.

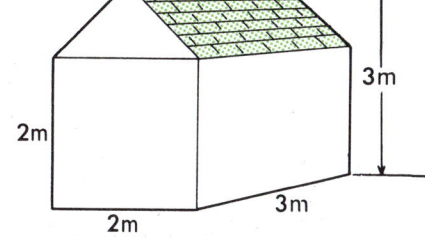

Work out:
**a.** the area of one wall

**b.** the area of one end

**c.** the total area to be painted.

**3. a.** Use a scale of 1 centimetre represents 1 metre to draw a plan of this field.

**b.** Work out the area of this field.

From your scale drawing find:

**c.** the angle of the corner where the tree is growing.

**d.** the perimeter of the field.

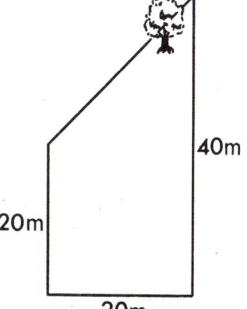

93

**Numbers from Ordered Pairs**

Do you remember learning about number lines like this one?

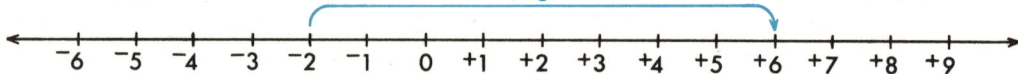

The line is a graph of the **set** of **integers**.
The blue arrow shows a movement from ⁻2 to ⁺6 which is in the + direction, and a move of 8 units, so a move from ⁻2 to ⁺6 is ⁺8.

This instruction can be given as an **ordered pair**.
**(⁺6, ⁻2) means move to ⁺6 from ⁻2.   (⁺6, ⁻2) = ⁺8.**

1.  Complete the following sentence:
    (⁻2, ⁺6) means move **to** . . . **from** . . . (⁻2, ⁺6) = ⁻ . . .

    Remember these directions: **positive → and negative ←.**

2.  Check that:   **a.**  (4, 8) = ⁻4    **b.**  (8, 4) = ⁺4.

●   When no + or − sign is shown, assume that it is +.

3..  **Starting at 0** write down the number at which you finish if you:

    **a.   move** ⁺1 unit then ⁻1 unit.     **b.   move** ⁺2 units then ⁻2 units.
    **c.   move** ⁻3 units then ⁺3 units.    **d.   move** ⁻2 units then ⁺2 units.

    In question **3** you were adding two integers but your starting and finishing points were 0.

●   ⁻2 is called the **opposite** of ⁺2 since ⁺2+ ⁻2 = 0
●   ⁺3 is the opposite of ⁻3, and ⁻3 is the opposite of ⁺3

4.  Write the missing numbers in:
    (4, 2) = ⁺ . . ., (5, 4) = . . ., (9, 6) = . . . and 2+1 = . . . so (4, 2)+(5, 4) = (9, 6)

    To add two integers which are represented by ordered pairs add the first numbers **and** add the second numbers.

    (4, 2)+(5, 4)          (7, 1)+(4, 6)
                     and
    = (9, 6)                = (11, 7)

    This number line shows a move of ⁺6 units followed by a move of ⁻2 units.

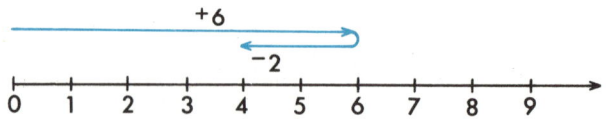

$$^+6 \ + \ ^-2 = \ ^+4$$

• To add a **negative integer** subtract its **opposite**.

$$^+6+^-2 = {}^+6-^+2 = {}^+4$$

5. Now check that $(7, 1)+(4, 6) = (11, 7)$ by completing the following:

    **a.**  $(7, 1) = {}^+...$    $(4, 6) = ...$    $(11, 7) = ...$

    **b.**  $^+6+^-2 = {}^+6-^+2 = 6-2 = ...$

## Let's Try

Use the number line on the opposite page to help you with these questions:

1. Find the replacement for each $\triangle$ in:

    **a.**  $4-6 = \triangle$    **b.**  $8-6 = \triangle$    **c.**  $5-10 = \triangle$    **d.**  $4-\triangle = {}^-3$

2. Write the integer which is represented by:  **a.**  $(4, 2)$    **b.**  $(3, 5)$    **c.**  $(7, 4)$

3. **a.** $(2, 0) = (3, 1) = {}^+2$.   Write four more integers which are equal to $^+2$.

    **b.** Write four more integers which are equal to $(0, 2)$.

4. Use the elements of $\{0, 1, 2, 3, 4\}$ to write:

    **a.** the two pairs which represent $^+3$.    **b.** the two pairs which represent $^-3$.

5. Write the pair $(0, ...)$ or $(..., 0)$ which represents the same integer as:

    **a.** $(1, 3)$ and $(8, 10)$    **b.** $(3, 1)$ and $(10, 8)$

6. Replace each $\triangle$ by a natural number.

    **a.** $(\triangle, 4) = {}^+2$    **b.** $(\triangle, 4) = {}^-2$    **c.** $(\triangle, 2) = 0$    **d.** $(7, \triangle) = {}^+6$

7. Fill in the missing part of:  **a.** $^+4+^-4 = ...$    **b.** $^-3+... = 0$

8. Replace the pairs by integers and find out if $(6, 4)+(1, 5) = (7, 9)$ is true.

9. Write an ordered pair which represents the same integer as:

    **a.** $(1, 4)+(2, 5)$    **b.** $(4, 2)+(2, 4)$    **c.** $(2, 0)+(1, 5)$

These packets of cereals have **net weights** of 450 grammes, and 250 grammes.
**Net weight** is the weight of the contents.

1. Complete this working to find the weight of cereals which can be bought for 1p.

   a. **Large-size packet**

   For 1p I can buy $\frac{450}{15}$ = . . . g

   b. **Small-size packet**

   For 1p I can buy $\frac{250}{10}$ = . . . g

   The working in question **1** shows that the large packet is the best buy, since I get 5 g more for each penny I spend.

2. Sometimes it is easier to find the **cost per gramme**. Complete the following working which shows you how to do this.

   a. **Large-size packet**
   450 g cost 15p

   so 1 g costs $\frac{15}{450}$ = $\frac{?p}{30}$

   = $\frac{?p}{300}$

   b. **Small-size packet**
   250 g cost 10p

   so 1 g costs $\frac{?}{?}$ = $\frac{1p}{25}$

   = $\frac{?p}{300}$

3. Which is the greater:

   a. $\frac{10}{300}$p, or $\frac{12}{300}$p?     b. $\frac{1}{30}$p, or $\frac{1}{25}$p?

   Not all our purchases are bought as **net** weights. These frozen peas look much more expensive than fresh ones. Notice that you are paying for **net** weight on the frozen peas whereas the weight of fresh peas includes the pods.

4. After shelling 1 kg of peas I weighed them and found:

   Weight of peas 600 g.
   Weight of pods 400 g.

   Compare the cost of the peas by completing this working:

   a. **Frozen Peas**

   For 1p I can buy $\frac{1000}{40}$ = . . . g

   b. **Fresh Peas**

   For 1p I can buy $\frac{600}{20}$ = . . . g

   c. Which is the best value for money: the frozen peas, or the fresh ones?

1. Work out the number of grammes which cost 1p if:

   a. 520 g cost 5p    b. $\frac{1}{2}$ kg costs 50p

   c. 450 g cost 8p    d. 675 g cost 15p

2. a. Calculate the volume of these pieces of nougat.

   b. What is the cost per gramme for each bar?

   c. Work out the number of grammes of each received for 1p.

   d. Write: the weight of B as a fraction of the weight of A, the cost of B as a fraction of the cost of A, and the volume of B as a fraction of the volume of A.

   e. Which is the best buy?

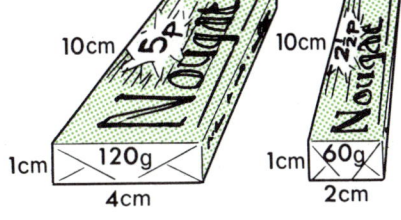

3. Work out which of these boxes of matches is the best buy.

4. Andrew is a philatelist; this means that he collects stamps. These packets of stamps are for sale and Andrew is wondering which is the best buy.

   a. What is the price per stamp in pack A?

   b. What is the price per stamp in pack B?

   c. Can Andrew be sure that pack A is the best buy without looking at the stamps?

5. These are some purchases from a grocer's shop. Find the number of grammes of each product received for 1p.

   a.

   b.

   c.

97

**Gradients**

1. Plot the following points on a pair of axes, and join each one to the origin.

   **a.** (2, 1)    **b.** (4, 1)

   **c.** (1, 1)    **d.** (6, 1)

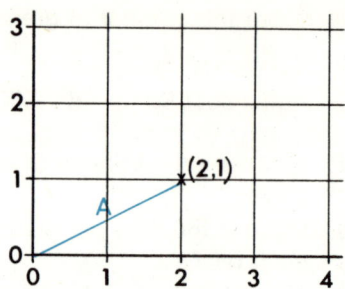

   Which is the steepest line on your graph? Which line is the least steep?

   The steepness of a line is written as a fraction. Line segment A has a slope of 1 in 2, or $\frac{1}{2}$.

2. **a.** Join (4, 2) to the origin on your graph. Has your new segment the same slope as line segment A?

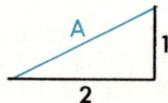

   | 2 in 4, or $\frac{2}{4}$ | has the same slope as 1 in 2, or $\frac{1}{2}$.

● The **slope**, or **gradient**, of your new line is 1 in 2. Use the fraction in its lowest terms to represent a gradient.

   For example, this line segment PQ has a gradient of

   $\frac{2\cdot5}{10} = \frac{1}{4}$ or

   $\frac{1\cdot5}{6} = \frac{1}{4}$

   The gradient of PQ is 1 in 4.

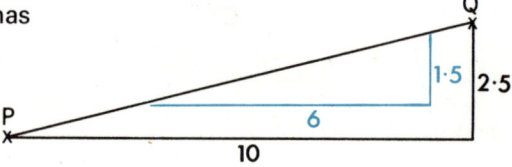

3. Here is a **road** gradient sign, and a **railway** gradient sign.

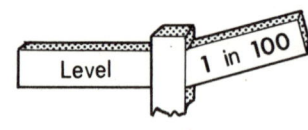

   **a.** Look for signs like these.

   **b.** 1 in 4 is quite a steep hill for a car to climb.
   Draw diagrams to show slopes of 1 in 4, and 1 in 10.

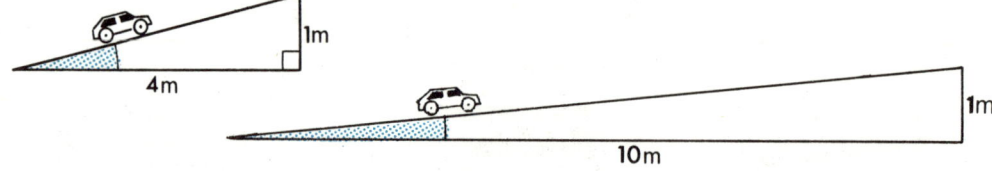

   Use a scale of 1 cm represents 1 metre and measure your angles shaded blue with a protractor.

1. **a.** Write the slope or gradient of these five line segments as a fraction in its lowest terms.

   **b.** Which line is the steepest?

   **c.** Which line is the least steep?

   **d.** Two of the lines have the same gradient. Which are they?

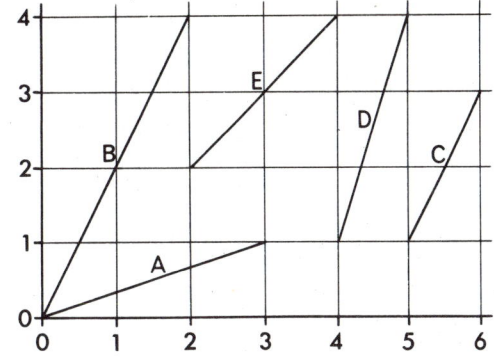

2. Using a 10-centimetre base line and a protractor, draw triangles which show gradients of:

   **a.** $\frac{3}{10}$  **b.** $\frac{1}{2}$  **c.** $\frac{1}{4}$  **d.** $\frac{3}{4}$  **e.** $\frac{2}{5}$

3. Draw axes from 0 to 8 using 1 centimetre to represent 1 unit.

   Through the origin draw lines with slopes:

   **a.** $\frac{1}{3}$  **b.** $\frac{2}{3}$  **c.** $\frac{3}{3}$  **d.** $\frac{4}{3}$

   Through (2, 3) draw, in a different colour, lines with slopes:

   **e.** $\frac{1}{3}$  **f.** $\frac{2}{3}$  **g.** $\frac{3}{3}$  **h.** $\frac{4}{3}$

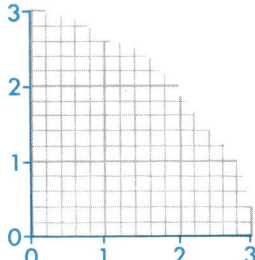

4. Mark a 1-metre long strip of card in 10-centimetre units. Use a 10-centimetre cube and a piece of wood to set up this apparatus.

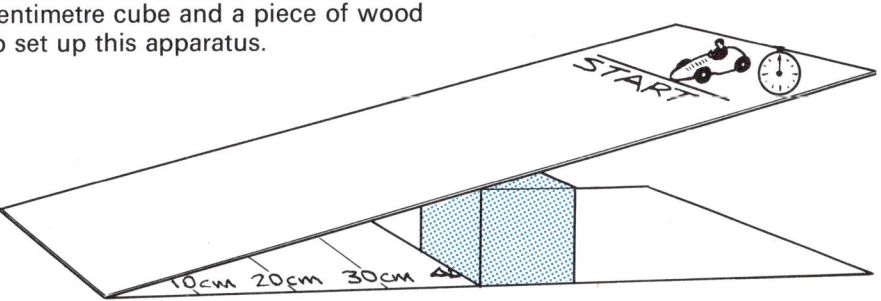

   You can find the time (seconds) taken for a car to roll down the slope.
   The gradient shown in the diagram is 10 in 40, or 1 in 4.
   Copy and complete this table and graph your results.

| Gradient | 1 in 1 | 1 in 2 | 1 in 3 | 1 in 4 |
|---|---|---|---|---|
| Time taken | | | | |

**The Percentage**

Jane, Debbie, and Susan were throwing
a die. They recorded the number of
sixes thrown in the following way:

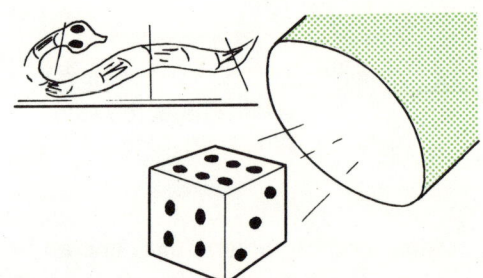

|  | Number of sixes |
|---|---|
| Jane | 2 out of 10 throws |
| Debbie | 5 out of 20 throws |
| Susan | 9 out of 50 throws |

1.  **a.**  $\frac{2}{10}$ of Jane's throws were sixes.
     Write the fraction of throws which were sixes for Debbie and Susan.
    **b.**  Complete this sentence.

     The fraction of throws which were sixes for Jane was $\frac{?}{100}$, for Debbie was $\frac{?}{100}$,

     and for Susan was $\frac{?}{100}$.

●  **When fractions have the same denominator it is easy to see which is the largest
and which is the smallest.**

Jane had this fraction: $\frac{2}{10}$, or $\frac{20}{100}$, or 20 per 100 which is 20 per cent.

●  **A fraction can be written as a percentage (%).**

Since $\frac{5}{20} = \frac{25}{100}$, 25% of Debbie's throws were sixes.
Since $\frac{9}{50} = \frac{18}{100}$, 18% of Susan's throws were sixes.

2.  **a.**  Which girl had the best percentage of throws?
    **b.**  Did Susan have the lowest percentage of throws?

To change a fraction to a percentage, first write the equivalent fraction with 100
as denominator.
For example:

$\frac{3}{5} = \frac{6}{10} = \frac{60}{100} = 60\%$

Here is a shop which is offering a
reduction in its prices.
The bicycle cost £30 before the sale.

3.  Complete this working:

    **a.**  15% of £30 $= \frac{?}{100} \times £30 = £\frac{?}{100} \times 30$

         $= £\dots$

    **b.**  The shopkeeper should reduce the
         cost of the bicycle by £4·50 and
         charge £30−£4·50 = . . .

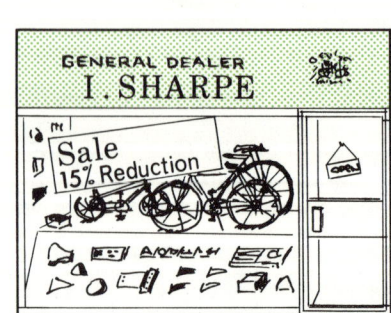

Percentages are often used in everyday life. Look out for the % sign in the shop windows.
Here is a bank which is offering 5% interest **per annum** (p.a.) which is per **year**.
A deposit of £100 will earn £5 interest, and a deposit of 100p will earn 5p interest.

TRUSTY BANK

5% Interest

4.  Complete this working.

A deposit of £8 will earn 5% of £8
interest $= \frac{5}{100} \times £8 = $ . . .p.
At the end of 1 year £8 becomes £8+. . .p = £8·40.

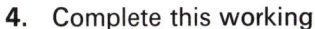

**Let's Try**

1.  Which is the best reduction in price?

Shop A  | $\frac{1}{5}$ **off all prices** |    Shop B  | **25% reduction** |

2.  A grocer finds that about 6% of his eggs are bad. How many are likely to be bad in a crate of:

   **a.**  100 eggs    **b.**  500 eggs    **c.**  1000 eggs

3.  A school has 440 children on roll. 45% of the children are girls.

   **a.**  What percentage of the children are boys?

   **b.**  How many boys, and how many girls are there in the school?

4.  **a.**  A leak in this oil tank causes a 10% loss. How many litres will be lost from a full tank?

   **b.**  In one week 30% of a full tank is sold. How many litres of oil are sold?

   **c.**  20% of a full tank is sold at 42p per litre and the rest is then sold at 43p per litre. How much does the garage receive for the oil?

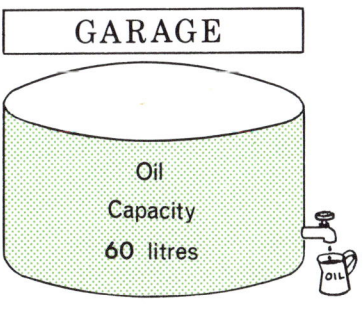

GARAGE

Oil
Capacity
60 litres

5.  A salesman is paid commission on his sales. He receives 15% of the money he collects and calls his share commission. Work out his commission on the following:

   **a.**  Week 1: £100    **b.**  Week 2: £150
   **c.**  Week 3: £160    **d.**  Week 4: £155

**Solutions from Venn Diagrams**

P.C. 49 tested the brakes and tyres on 40 bicycles. This is a Venn diagram showing:

C = {Bicycles}
T = {Bicycles with faulty tyres}
B = {Bicycles with faulty brakes}

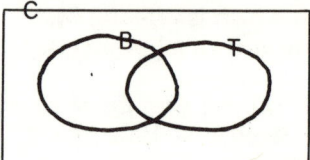

The number of elements in the **universal set**, C, is 40.

P.C. 49 found that:

(1) bicycles with **faulty brakes** but **good tyres** numbered 12.

(2) bicycles with **faulty tyres** but **good brakes** numbered 8.

(3) bicycles with no faults numbered 6.

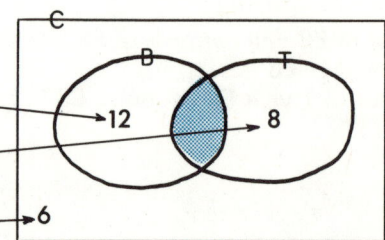

1. Write the missing number after checking on the Venn diagram.

   a. The **blue** region contains bicycles with faulty tyres and faulty brakes. The number in this region is $40 - (12+6+8) = \ldots$

   b. Altogether $14+12 = \ldots$ bicycles had faulty brakes.

   c. $14 + \ldots = \ldots$ bicycles had faulty tyres.

   d. The number of faulty bicycles was $12 + \ldots + 8 = 34$
      or $40 - \ldots = 34$

● Venn diagrams can be used to represent information and help us solve problems.

For example:

In my wardrobe I have 24 patterned ties. 4 of them are blue, but not green, 6 of them are blue and green, and 7 of them are neither blue nor green. Notice that this information has been shown on the diagram.

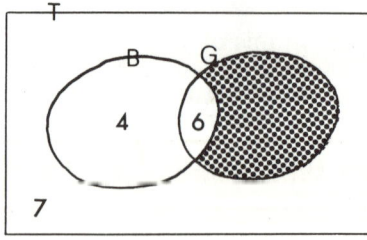

T = {Ties}
B = {Blue ties}
G = {Green ties}

2. a. Write the missing number in the shaded region which shows the number of ties which are **green but not blue**.

   b. How many ties are green? Remember to include ties which are a mixture of green and blue.

   c. The number of ties which are **not green** is $4+7$. How many ties are not blue?

**1.** C is the set of children in a class.
G = {girls} and L = {left-handed children}
There are 30 children in the class: n(C) = 30.

Write the number of:

**a.** left-handed boys

**b.** left-handed children

**c.** left-handed girls

**d.** girls

**e.** boys

**f.** children who are either left-handed
or girls

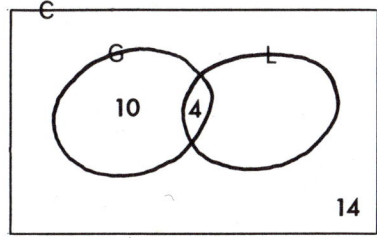

**2.** C = {children} and n(C) = 30.
R = {children with a pet rabbit}
D = {children with a pet dog}

Copy this Venn diagram and enter the
following information.

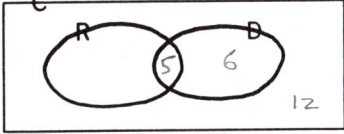

**a.** 12 children keep neither a rabbit nor a dog as a pet.

**b.** 6 children keep a dog but not a rabbit.

**c.** 5 children keep a dog and a rabbit.

How many children keep:

**d.** a rabbit only?     **e.** a rabbit?

**f.** a dog?              **g.** at least one pet?

**3.** L is the set of natural numbers less
than 20.

List the subsets, S and R, of L where:

**a.** S = {square numbers}

**b.** R = {rectangular numbers}

**c.** Copy this Venn diagram and show
the elements of L on it.

**d.** Draw a new Venn diagram and
enter:
> **the number** of rectangular
> numbers which are not square
> **the number** of square
> numbers

**e.** From your Venn diagram work out
the number of elements of L which
are neither square nor rectangular.

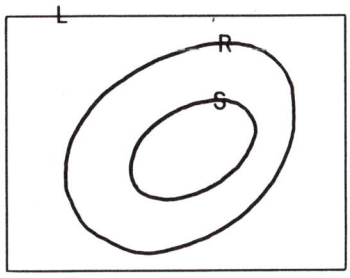

1. **a.** Cut a rectangle of card 10 cm by 5 cm.

   **b.** Find the centre of symmetry X by drawing the two dotted lines.

   **c.** On the clean side of your rectangle, write your initials in one corner. Now pin your rectangle through the centre of symmetry and draw a thick green line round your rectangle.

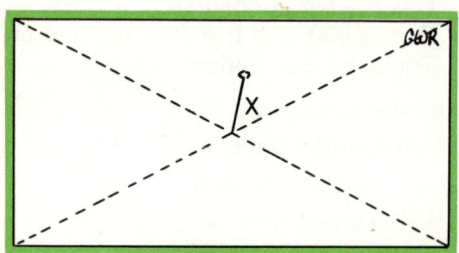

- Each of the dotted lines joins opposite vertices of the rectangle and is called a **diagonal**.
- The diagonals intersect at the centre of symmetry.

Your rectangle can be rotated about the centre of symmetry.

2. **a.** Rotate your card clockwise until it fits in the green rectangle again.

   **b.** Now rotate clockwise again until the card fits in the rectangle. Is the card in its starting position?

To fit the rectangle in its green "frame" we can rotate:

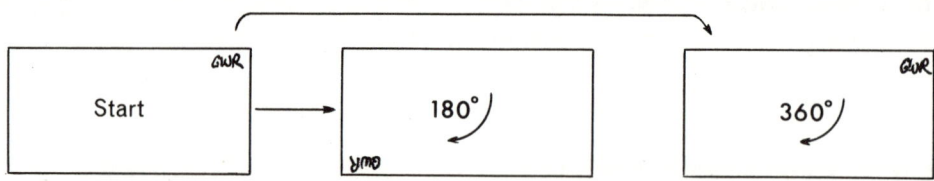

3. **a.** Cut out an **equilateral triangle** from card.

   **b.** Locate its centre of symmetry by folding along the **lines** of **symmetry**.

   **c.** Put your initials in one corner and pin through the centre of symmetry.

   **d.** Draw a frame round the triangle. In how many ways will the triangle fit into its frame?

   **e.** How many lines of symmetry has an equilateral triangle?

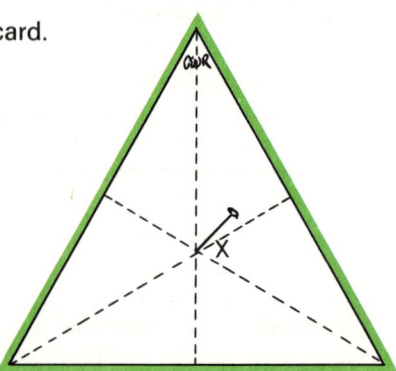

To fit an equilateral triangle into its frame we can rotate it:

 Start 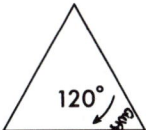 120°  240°  360°

Remember that 1 complete revolution is equal to 360 degrees, or 360°.

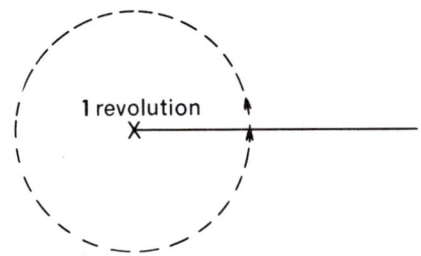

1 revolution

**4.** Complete these statements.

   **a.** $180° \times 2 = \ldots° = \ldots$ revolution

   **b.** $120° \times 3 = \ldots° = \ldots$ revolution

**1.** **a.** Draw and cut out a 10-cm square. Locate its centre of symmetry and mark one corner.

   **b.** Pin through the centre of symmetry and draw a frame round the square.

   **c.** How many lines of symmetry has a square?

   **d.** In how many ways will the square fit in the frame?

   **e.** Write the four angles through which you can rotate your square in order to fit it in the frame.

X

**2.** Repeat question **1** with a:

   A   regular hexagon
   B   parallelogram

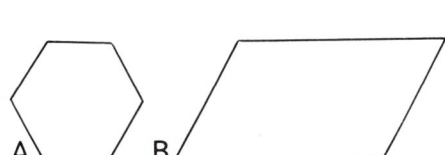

A     B

**3.** Use your equilateral triangle for this question.

$R_{120} + R_{240}$ means

"Rotate 120° after rotating 240°".

   **a.** Check that $R_{120} + R_{240}$ gives the same result as $R_{360}$, that is $R_{120} + R_{240} = R_{360}$

Write the missing instruction using $\{R_{120} \ R_{240} \ R_{360}\}$.

   **b.** $R_{240} + R_{120} = \ldots$    **c.** $R_{120} + R_{120} = \ldots$

   **d.** $R_{360} + R_{120} = \ldots$    **e.** $R_{360} + \ldots = R_{240}$

## Weight per Cubic Centimetre

Here is a box of beef cubes which I bought from the supermarket.

The box has **net weight** 60 g and contains 10 cubes of beef extract.

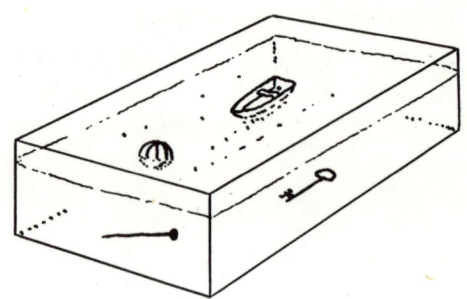

1. What is the weight of each cube?

2. Each cube is 2 cm by 2 cm by 2 cm. Write the volume of one cube in **cubic centimetres**.

   The weight of 8 cubic centimetres of beef extract is 6 grammes, so 1 cubic centimetre weighs $\frac{6}{8} = \frac{3}{4} = 0.75$ g.

   It is easy to see that iron is heavier than wood, but is cork heavier or lighter than wood?

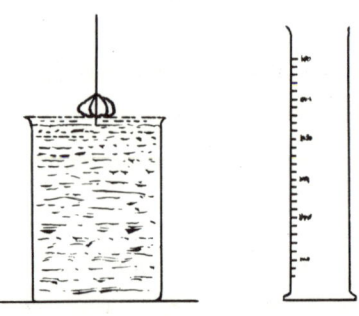

● In order to decide whether one substance is heavier than another we need to compare weight with volume and find: **weight per unit of volume**.

Remember that the weight of 1 cubic centimetre of water is 1 gramme, so the weight of 1 cubic centimetre of a substance which floats in water will be less than 1 gramme, and the weight of 1 cubic centimetre of a substance which sinks in water will be greater than 1 gramme.

3. Find out which of these substances float:
   rubber, balsa wood, plastic, polystyrene, oil, and sand.
   Take care with your experiment. An empty plastic bottle will float because it is full of air. Does this prove that plastic floats?

4. Follow these instructions which show you a method of finding the volume of an awkward-shaped solid, such as a stone.

   a. Fill a beaker to the brim with water.

   b. Place the beaker on a tray and then gently lower the stone into the water. Collect the water which overflows in the tray.

   c. Measure the volume of this water in a measuring cylinder.

The volume of the stone equals the volume of water collected.

Suppose that the volume of the stone is 50 cubic centimetres and the weight is 175 grammes, then the weight of 1 cubic centimetre is $\frac{175}{50}$ = 3·5 grammes.

**5.** How many times heavier than water is the stone?

Let's Try

**1.** Work out the weight per cubic centimetre of these solids:

**a.**
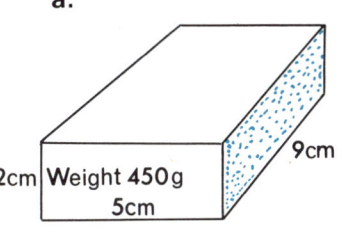
2cm | Weight 450g
5cm
9cm

**b.**
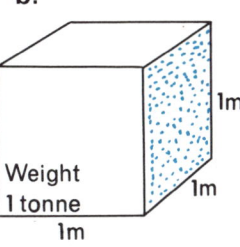
Weight 1 tonne
1m
1m
1m

**c.**
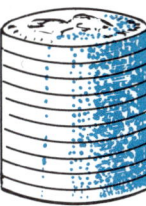
10 one penny coins

Use the water method to find the volume

**2.** Here are some sticks of chalk.

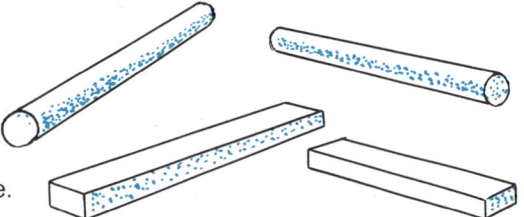

   **a.** Why can't you find the volume of chalk by putting it in water? Try it.

   **b.** Using a rectangular stick of chalk find its weight per cubic centimetre.

**3.** Find the weight per cubic centimetre of carrot and turnip. You can either carve out a cuboid of each or use the water method of finding volume.

**4.** There was once a king who gave his goldsmith 10 kilogrammes of gold and ordered him to turn it into a crown. The king thought that he had been swindled even though the crown weighed 10 kilogrammes. He suspected the goldsmith of mixing some cheaper metal with the gold. The goldsmith was eventually beheaded after the king found that his crown did not have the same volume as 10 kilogrammes of gold. Can you think of the way in which the king discovered the deception?

**5.** "John the tailor makes suits from 500-g and 750-g cloth."

   **a.** Find out what is meant by the weight of cloth.

   **b.** Choose a piece of material and find its weight.

Here are some items from a
bricklayer's tool bag. He uses
the trowel to lay bricks, and
the string to help keep the
bricks in line.

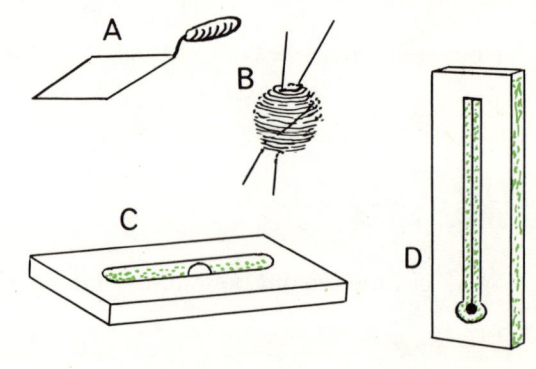

Item C is a **spirit-level** which
is used to check for **horizontal**.
Your classroom floor is
horizontal.

Item D is a **plumb line** and is
used to test that walls are
**vertical**.

1.  **a.** The line segment where your
    classroom wall intersects with the
    floor is **horizontal**. Make a list of
    some more horizontal line
    segments in your classroom.

    **b.** The edge of the door-frame is
    **vertical**. Make a list of some
    more vertical line segments. Do
    not forget the flag-pole.

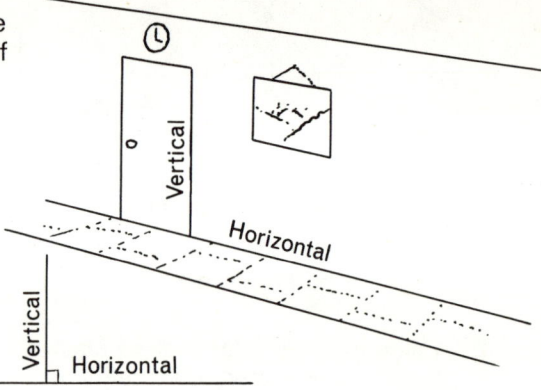

● The angle between vertical and
horizontal is 1 right angle, or 90°.

2.  Make a list of some things or line
    segments which are neither vertical
    nor horizontal. Do not forget the
    Leaning Tower of Pisa.

    This diagram shows an angle measurer
    which you can make.

    Mark a quarter of a circle in degrees,
    fasten a milk straw onto the edge as a
    sighting tube, and then hang a plumb
    line. Make sure that the tube is on the
    correct edge and the line is fastened
    at the correct point.

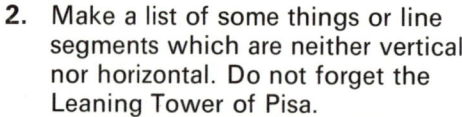

3.  This picture shows a boy using his
    angle measurer which reads 40° when
    he is 50 metres from the foot of the
    flag-pole.

    The **angle of elevation**
    is 40°.

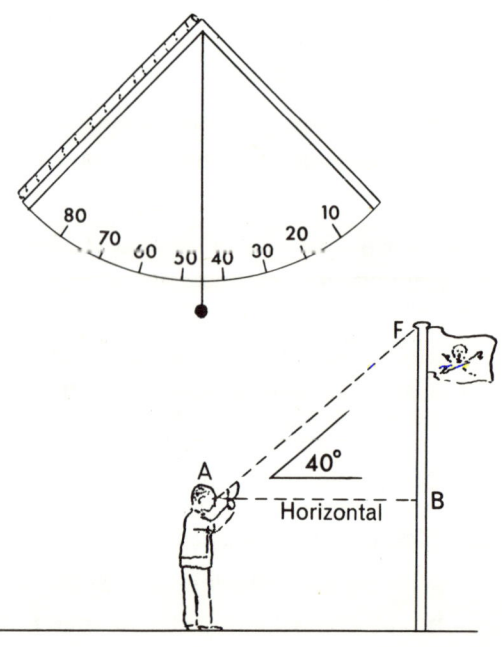

**a.** Draw a line segment $\overline{AB}$, 50 cm long, near the bottom edge of a sheet of paper.

**b.** Measure an angle of 40° at A, and a right angle at B.
Check that m($\overline{FB}$) is about 42 cm.

The scale you have used is 1 cm represents 1 metre, so 42 cm represents 42 metres.

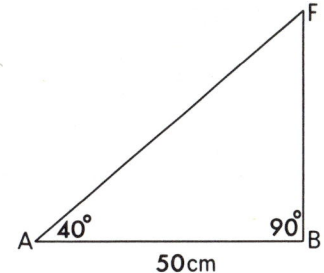

* The flag-pole is 42 metres **plus** the boy's **eye-level height**.

**c.** Complete this sentence:
The boy's eye-level height is 1·4 metres, so the height of the flag-pole is 42+1·4 = . . . metres.

Let's Try

**1.** With the help of a friend, measure your own eye-level height (h cm).

**2.** Find a bottle with a screw top and fill it almost full of water. You can use this bottle as a spirit-level. Notice how the bubble moves. Check some surfaces which you think are horizontal.

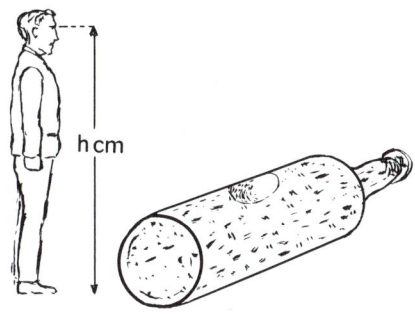

**3.** This diagram shows John measuring the height of a factory chimney.

**a.** Which line segment is parallel to $\overline{AB}$?

**b.** Which two line segments are horizontal?

**c.** Write a line segment which is vertical?

**d.** John is 30 metres from the chimney and finds that the angle of elevation is 60°. Use a scale of 1 cm represents 1 metre to find the height of the chimney. John's eye-level height is 1·5 metres.

**4.** Use your angle measurer to find the height of:

**a.** a telegraph pole **b.** your school

**5.** Find out about a **theodolite** which is used by a **surveyor**.

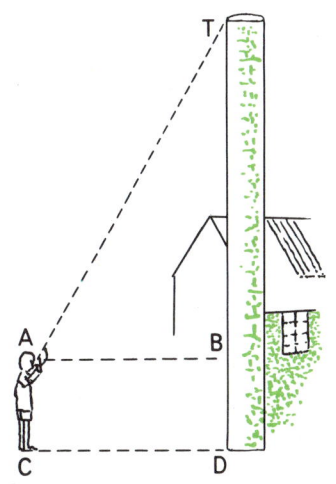

109

Andrew has received a new bicycle
for his birthday and is playing with
Barry and Chris. All three boys wish to
have a ride but cannot decide who
should be first. The order could be
A, B, C, that is Andrew then Barry
and then Chris.

1.  Complete all the possible orders:

    A B C    A – –    B A –
    B – –    C A –    C – –

● There are 6 "orders" for the three boys. Each order of the three letters is called a
**permutation**.

2.  Chris decides to go home leaving Andrew and Barry to play with the bicycle.
    Write the two permutations of "A" and "B".

I have three number cards which must
be used to fill this hymn number
board.

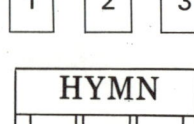

Either 1, or 2, or 3 can be put in the
first space.

3.  How many numbers are then left to
    put:
    in the second space, and then in the
    third space?

Having chosen one of **three** numbers for the first space, I can then choose one
of **two** numbers for the second space, and then have **one** number left for the
third space.

The number of ways of arranging the three cards ⟨1⟩ ⟨2⟩ ⟨3⟩ is
**three**×**two**×**one** = 6

4.  a.  Write the six possible hymn numbers using all of ⟨1⟩ ⟨2⟩ ⟨3⟩

    b.  Compare part **a** with question 1. Do you notice the similarity?

With only two cards ⟨3⟩ ⟨2⟩ there are two ways of choosing the first number,

and **one** way for the second.

 and ⟨2⟩⟨3⟩

5.  Compare the arrangements of ⟨3⟩ and ⟨2⟩ with question **2**.

    The number of ways of arranging two cards   is
    **two**×**one** = 2

**6.** Derek has now joined Andrew, Barry, and Chris. Complete this list of permutations for the order in which he will ride Andrew's bicycle.

```
A B C D     A B D C     A C – –     A C – –     A D – –     A D – –
B A C D     B A D C     B C – –     B C – –     B D – –     B D – –
C A B D     C A D B     C B – –     C B – –     C D – –     C D – –
D A B C     D A C B     D B – –     D B – –     D C – –     D C – –
```

Number of orders = 4×3×2×1 = . . .

Mathematicians have invented a short way of writing 4×3×2×1.

4×3×2×1 = 4! which is read as **factorial four**.

The number of permutations of two symbols 2! or 2×1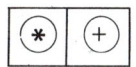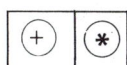

● The number of permutations of three symbols is factorial three, 3!

## Let's Try

**1.** You have probably recited a rhyme which ends, "out goes she".
Write all the different orders of the words, for example: "she goes out".

**2.** Here are five books from my bookshelf. Use the letters to list all the different arrangements of the books. Are there 5! ways of arranging the books?

**3.** Here is a plate of five cakes.
In how many ways can you choose:

  **a.** the first one?

  **b.** the second one?

  **c.** the third one?

  **d.** the fourth one?

  **e.** the fifth one?

  **f.** In how many different orders can you eat five cakes?

**4.** Copy and complete this table of factorials.

| 1! | 2! | 3! | 4! | 5! | 6! | 7! | 8! |
|----|----|----|----|----|----|----|----|
|    |    |    |    |    |    |    |    |

**5.** Seven girls have been chosen to represent the school at netball. If each player can play in any position, in how many ways can the team be arranged?

**The Chance**

1. Copy and complete this table for
   left-handed (L) and right-handed (R)
   children in our class:

   L RRRR L RRR LL RRRR L RRRR
   R L RRR L RRRR LL RRRRRR L R

|  | Tally | Frequency |
|---|---|---|
| Left |  |  |
| Right |  |  |
| **Total** |  | 40 |

2. Write the missing numbers and words in the following sentences:

   a. The number of left-handed children is . . .
   b. $\frac{1}{4}$ of the children are . . . handed.
   c. The number of right-handed children is . . .
   d. $\frac{?}{?}$ of the children are right-handed.

   The teacher asked a stranger to guess whether the class monitor is left-handed or right-handed.

3. Is the stranger more likely to be correct if he says, "Right", than if he says, "Left"?
   For our class:

   The **chance** that a pupil is right-handed is 3 in 4, or $\frac{3}{4}$.
   The **chance** that a pupil is left-handed is 1 in 4 or $\frac{1}{4}$.

   **Who goes first?** You will probably toss a coin and call "Head", or "Tail".
   The coin can fall two ways, Head (H)                    or Tail (T)

   The **chance** of falling "head" is
   1 head out of 2 ways,
   that is 1 in 2, or $\frac{1}{2}$.
   The **chance** of falling "tail" is 1 tail out of
   2 ways, that is 1 in 2, or $\frac{1}{2}$.

   The chance that a coin will fall "head" is equal to the chance that it will fall "tail".

4. Toss a coin twice. You might get one "head" and one "tail".

   Since the chances are both $\frac{1}{2}$ you would expect to get HTHTHTH . . . but this does not happen in practice.
   If you toss the coin many times there will be about $\frac{1}{2}$ heads and about $\frac{1}{2}$ tails.

5. In 200 tosses I tallied 97 "heads" and 103 "tails".

Which of these fractions: $\frac{103}{200}$ or $\frac{97}{200}$, gives me:

**a.** the fraction of heads?    **b.** the fraction of tails?

**c.** Is 97 **about** $\frac{1}{2}$ of 200?    **d.** Is 103 **about** $\frac{1}{2}$ of 200?

Let's Try

**1.** Toss a coin 100 times.
Ask four friends to do the same and
record your results so that you have a
total of 500 tosses.

Use your results to check that in
tossing a coin the chance of "heads"
is $\frac{1}{2}$, and "tails" is $\frac{1}{2}$.

|  | Heads | Tails |
|---|---|---|
| 1 | | |
| 2 | | |
| 3 | | |
| 4 | | |
| 5 | | |
| **Totals** | | |

**2.** Toss a button and find the chance that it will fall the right way up and the chance that it will fall upside down.

**3.** What is the chance of choosing **red**, or **green**, from the two colours.
Test your answer by interviewing fellow pupils. Do not be surprised at your result; some colours are more popular than others.

**4.** A die can fall in one of six ways: 1, 2, 3, 4, 5, and 6.
Write the fractions to show the chance of getting

**a.** 1    **b.** 2    **c.** 3    **d.** 4    **e.** 5    **f.** 6

Test your answers by throwing a die a large number of times.

**5.** Find out which is the most popular number from {1, 2, 3, 4, 5} by interviewing fellow pupils.
What is the chance of choosing:

**a.** 1    **b.** 2    **c.** 3    **d.** 4    **e.** 5

**6.** After interviewing 1000 housewives it was found that 650 used **Sudso**, and 350 used **Frothy**.

The fraction of housewives using:

**a.** Sudso is . . .    **b.** Frothy is . . .

The chance that the next housewife to be interviewed will use:

**c.** Sudso is . . .    **d.** Frothy is . . .

Out of 10000 housewives how many do you think will use:

**e.** Sudso    **f.** Frothy

**7.** Find out what a **census** and a **mini-census** are.

# Answers

**Page 7**
1. **a.** OXO, TOT   **b.** DID, BOB
2. **a.** Two   **b.** Two   **c.** Many ways.

3. **b.**

| 1 | 4 | 8 | 4 |
|---|---|---|---|
| 2 | 2 | 4 | 2 |
| 3 | 3 | 6 | 3 |
| 4 | 5 | 10 | 5 |

4. **a.**                **b.**                **c.**                **d.**

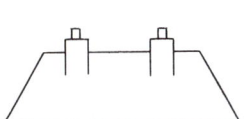      

**Page 9**
1. **a.**

| A | Yes | Yes | Yes |
|---|-----|-----|-----|
| B | Yes | Yes | Yes |
| C | Yes | No  | No  |
| D | No  | Yes | No  |

**b.** A  4
   B 16
   C  5
   D  2

2. **a.**

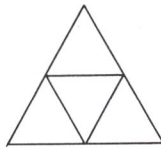

**b.**

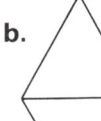

**Page 11**
1. **a.** A Square; B Rectangle; C Rhombus; D Parallelogram; E Trapezium
   **b.** E   **c.** A, B, C, D
2. **a.** 2·5 cm          **b.**                **c.**

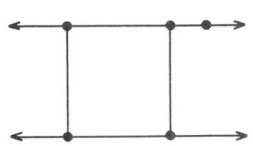      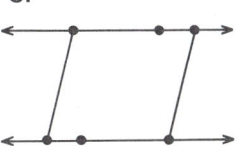

3. **a.** 52° North
5. **b.** Three
6. Eighteen

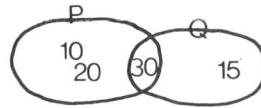

**Page 13**
1. **a.** {10, 20, 30}   **b.** {15, 30}   **c.** {10, 20, 30, 15}   **d.** {30}   **e.** {10, 20}
   **f.** {15}
2. **a.** {Mary, Tom, Andrew, Jane, Bill, Ann}   **b.** {Steven, John, Jane, Bill, Ann}
   **c.** {Jane, Bill, Ann}   **d.** {Mary, Tom, Andrew, Jane, Bill, Ann, Steven, John}
   **e.** {Mary, Tom, Andrew}   **f.** {Steven, John}
3. **b.** A ∪ V = A; A ∩ V = V   **c.** { } The set is empty.

**Page 15**

1. **a.** {1, 2, 7, 14}   **b.** {1, 2, 4, 7, 14, 28}   **c.** {1, 2, 5, 10, 25, 50}
   **d.** {1, 2, 4, 5, 10, 20, 25, 50, 100}
2. **a.** {2}   **b.** {3}   **c.** {2, 3, 5}   **d.** {2, 5}   **e.** {43}   **f.** 43
3. Highest common factor:
   **a.** 5   **b.** 9   **c.** 25   **d.** 1   **e.** 1   **f.** 37   **g.** 100   **h.** 43
4. **a.** 658   **b.** 1936   **c.** 7170   **d.** 41   **e.** 41   **f.** 78
5. 101 and 103
6. Yes

**Page 17**

1.

| Age | 9 | 10 | 11 | 12 | |
|---|---|---|---|---|---|
| Number | 10 | 13 | 18 | 12 | **Total** 53 |

3. **a.** Yes   **b.** Yes   **c.** No. More children should have been interviewed.

**Page 19**

1. **a.** $3\frac{4}{5}$   **b.** $4\frac{1}{5}$   **c.** $2\frac{3}{20}$   **d.** $3\frac{17}{20}$   **e.** $2\frac{3}{20}$   **f.** $5\frac{7}{20}$
2. **a.** $1\frac{1}{5}$   **b.** $1\frac{4}{5}$   **c.** $1\frac{7}{10}$   **d.** $4\frac{13}{20}$   **e.** $\frac{17}{20}$   **f.** $2\frac{7}{20}$
3. **a.** $1\frac{9}{20}$ m   **b.** $1\frac{1}{10}$ m   **c.** $1\frac{3}{20}$ m   **d.** $1\frac{17}{20}$ m
4. **a.** $\frac{1}{2}$ kg and $\frac{3}{5}$ kg   **b.** $1\frac{1}{10}$ kg
5. **a.** $1\frac{1}{5}$   **b.** $2\frac{1}{2}$

**Page 21**

1. **a.** 272 km   **b.** 294 km   **c.** 38 km   **d.** 314 km
2. **a.** Manchester   **b.** York   **c.** Northampton   **d.** York
3. **a.** Knotty                     **b.** 2 hours; $2\frac{1}{2}$ hours

| 55 | Buzztown | | |
|---|---|---|---|
| 75 | 50 | Lexton | |
| 120 | 95 | 45 | Wick |

4. 176 km

**Page 23**

1. **a.** 13   **b.** 26   **c.** 27   **d.** 87
2. **a.** $11110_{(2)} = 30_{(10)}$   **b.** $10101_{(2)} = 21_{(10)}$   **c.** $1101_{(2)} = 13_{(10)}$
   **d.** $11111_{(2)} = 31_{(10)}$
3. **a.** "One nought one nought" $1010_{(2)}$   **e.** "One one one" $111_{(2)}$
4. **a.** $100000_{(2)}$   **b.** $110000_{(2)}$   **c.** $1100000_{(2)}$   **d.** $1110000_{(2)}$   **e.** $10000_{(2)}$
   **f.** $100000_{(2)}$   **g.** $11000_{(2)}$   **h.** $1100100_{(2)}$

**Page 25**

2. **a.** 25, 29, . . .   **b.** 25, 30, . . .   **c.** 26, 30, . . .   **d.** 10, 15, . . .
3. **b.** Square
5. "Differences" form the same series.

**Page 27**

1. **a.** 1 cm represents 50p.   **b.** 1 cm represents 5 francs.   **c.** 6; 18; 15
   **d.** 75; 175; about 83
2. **a.** 21p   **c.** Total = 15p   **d.** 231p or £2·31
3. **a.** 15 min   **c.** 70 cm; 75 cm; 82·5 cm   **d.** $5\frac{1}{2}$ min; $5\frac{1}{4}$ min; 12 min

**Page 29**

1. **a.** No  **b.** Yes
2. **a.** 5, 10, 15, 20, 25, 30, 35, 40, 45, 50   **b.** The digit in the units column is 0 or 5.
3. **a.** Even  **b.** 10
4. (3, 5); (5, 7); (11, 13); (17, 19); (29, 31); (41, 43); (59, 61); (71, 73)
5. **a.**  36  **b.** Complete factorization: $2 \times 2 \times 3 \times 3$
   $$4 \times 9$$
   Set of prime factors:   {1, 2, 3}
   $$2 \times 2 \times 3 \times 3$$
6. **a.** $1+2+4+7+14 = 28$  **b.** $1+2+3 = 6$; $1+2+4+8+16+31+62+124+248 = 496$
7. **a.** {1, 2, 5}  **b.** {1, 2, 5}  **c.** {1, 2, 5}  **d.** {1, 2, 97}

**Page 31**

1. **a.** 0·13  **b.** 0·35  **c.** 0·95  **d.** 0·16  **e.** 0·48  **f.** 0·18  **g.** 0·025  **h.** 0·075
   **i.** 0·125  **j.** 0·875
2. **a.** 0·75  **b.** 5 cm; 0·05 m  **c.** No
3. **a.** 0·1̇  **b.** 0·2̇  **c.** 0·3̇  **d.** 0·4̇  **e.** 0·5̇  **f.** 0·16̇  **g.** 0·83̇  **h.** 0·25̇
4. **a.** <  **b.** >  **c.** =
5. 0·14285714 . . . It never ends.

**Page 33**

1. **a.** 18 min  **b.** 19½ min  **c.** 45 min  **d.** 67½ min  **e.** 60 km  **f.** 32 km
   **g.** 46⅔ km  **h.** 66⅔ km
2. **a.** 2 km  **b.** 3·5 km  **c.** 5·1 km
3. **a.** 1 cm represents 30 min.  **b.** 1 cm represents 5 km.  **c.** 20 km per hour
   **e.** 15 km, 25 km, 44 km

**Page 35**

2. **a.** 0·2  **b.** 0·35  **c.** 0·55  **d.** No
3. **a.** 2  **b.** 10  **c.** 100  **d.** 1000
4. **a.** 0·1  **b.** 0·02  **c.** 0·765  **d.** 0·001
5. **a.** 0·4  **b.** $\frac{1}{15}$  **c.** $\frac{14}{15}$  **d.** 550 g  **e.** 2000

**Page 37**

1. **a.** 11  **b.** 7  **c.** 1  **d.** 18
2. **a.**

| Total | 30 | 31 | 23 |
|---|---|---|---|
| Average | 7·5 | 7·75 | 5·75 |

   **b.** Tickler  **c.** Waller  **d.** Total 84 runs; average 7 runs per innings
3. **a.** £12·15  **b.** £2·43
4. **a.** 1890 litres  **b.** 210 litres
5. 22½p

**Page 39**

1. **a.** 100  **b.** 25  **c.** 12·5  **d.** 4  **e.** 3·5
2. **a.** 2  **b.** 1·2  **c.** 10·5  **d.** 0·5  **e.** 0·9
4. **a.** $\frac{1}{100}$  **b.** $\frac{1}{10}$  **c.** 1

**Page 41**
1. **a.** £0·02  **b.** £0·15  **c.** £0·40  **d.** £1·20  **e.** £1·02  **f.** £2·55  **g.** £1·99
   **h.** £9·19
2. **a.** £$\frac{1}{100}$  **b.** £$\frac{3}{20}$  **c.** £$\frac{19}{40}$  **d.** £$\frac{1}{2}$  **e.** £$\frac{21}{40}$  **f.** £$\frac{3}{4}$  **g.** £$\frac{7}{8}$  **h.** £$\frac{199}{200}$
3. **a.** 30p  **b.** 60p  **c.** 35p  **d.** 17$\frac{1}{2}$p
4. **a.** 140p or £1·40  **b.** 120p or £1·20  **c.** 190p or £1·90  **d.** 55p

5. 

| | | |
|---|---|---|
| $\frac{1}{10}$ | 35p | £3·15 |
| $\frac{1}{5}$ | 70p | £2·80 |
| $\frac{1}{4}$ | 87$\frac{1}{2}$p | £2·62$\frac{1}{2}$ |
| $\frac{7}{10}$ | £2·45 | £1·05 |

6. **a.** 19  **b.** $\frac{1}{3}$  **c.** $\frac{2}{3}$

**Page 43**
1. **a.** $\frac{3}{8}$  **b.** $\frac{7}{20}$  **c.** $\frac{1}{10}$  **d.** $\frac{1}{8}$  **e.** $\frac{2}{5}$  **f.** $\frac{7}{10}$
2. **a.** $\frac{9}{20}$  **b.** $\frac{3}{10}$  **c.** $\frac{3}{25}$  **d.** $\frac{1}{8}$  **e.** $\frac{21}{100}$  **f.** $\frac{8}{5} = 1\frac{3}{5}$
3. **a.** 10  **b.** $\frac{3}{5}$  **c.** 10 (Tuesday), 5 (Wednesday)
4. **b.** $\frac{3}{11}$  **c.** 6 m²  **d.** $\frac{1}{16}$ m²  **e.** 96 blue area  256 grey area
5. **a.** 60 degrees  **b.** 600 g  **c.** 700 m  **d.** 40p

**Page 45**
1. **a.** {(1, 24), (2, 12), (3, 8), (4, 6), (6, 4), (8, 3), (12, 2), (24, 1)}
   **b.**
   
   Points lie on a curve
   
   **c.**
   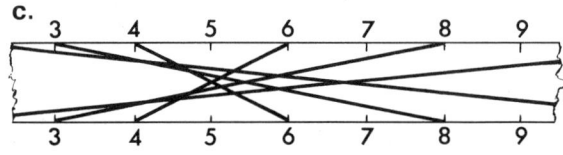

2. **a.**

| 1 | 2 | 3 | 4 | 5 | 6 |
|---|---|---|---|---|---|
| 3 | 6 | 9 | 12 | 15 | 18 |

   **b.**
   
   Points lie on a straight line
   
   **c.**
   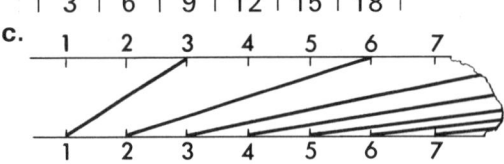
   
   **d.** 7$\frac{1}{2}$ km
3. **a.**

| L | 1 | 2 | 3 | 4 | 5 | 6 | 7 | 8 | 9 | 10 | 11 |
|---|---|---|---|---|---|---|---|---|---|----|----|
| B | 11 | 10 | 9 | 8 | 7 | 6 | 5 | 4 | 3 | 2 | 1 |
| A | 11 | 20 | 27 | 32 | 35 | 36 | 35 | 32 | 27 | 20 | 11 |

   **b.**
   
   Points lie on a curve
   
   **c.**
   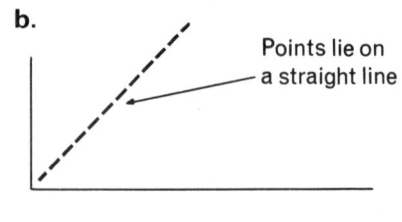
   
   **d.** L = 6 m
   B = 6 m

4. **a.** {(0, 0), (2, 1), (4, 2), (6, 3), (8, 4), (10, 5)}
   **b.** {(10, 0), (9, 1), (8, 2), (7, 3), (6, 4), (5, 5), (4, 6), (3, 7), (2, 8), (1, 9), (0, 10)}
   **c.** {(0, 0), (3, 1), (6, 2), (9, 3)}

**Page 47**
1. a. $100_{(2)}$   b. $1100_{(2)}$   c. $1100_{(2)}$
2. a. $11111_{(2)}$   b. $100110_{(2)}$   c. $101001_{(2)}$   d. $101000_{(2)}$   e. $101010_{(2)}$
   f. $1001011_{(2)}$
3. a. $10_{(2)}$   b. 1   c. $11_{(2)}$
4. a. $1000_{(2)}$   b. $1010_{(2)}$   c. $1001_{(2)}$   d. 1   e. $1101_{(2)}$   f. $100110_{(2)}$
5. Sum          a. 1101   b. 1110   c. 100000
   Difference      1          100          10
                                   (Base two)
6. $11111_{(2)}$, $11111_{(2)}$, $11110_{(2)}$, Total: $1011100_{(2)}$

**Page 49**
1. a. 10000   b. 9200   c. 750   d. 80
2. a. 2   b. 0·21   c. 0·907   d. 0·05
3. a. 54   b. $16\frac{4}{5}$   c. 600   d. 900
4. a. $\frac{1}{2}$   b. $\frac{1}{4}$   c. $\frac{4}{15}$   d. $\frac{31}{500}$
5. a. 130 kg   b. $\frac{92}{105}$   c. $\frac{13}{105}$   d. 0·92 and 1·05 tonnes
6. a. 12; 8   b. $\frac{5}{8}$, 7500   c. 60 m³   d. Yes; yes; no

**Page 51**
1. a. B   b. A, C, E
2. b. 6   c. 12
3. a. 20, 20, 12
5.

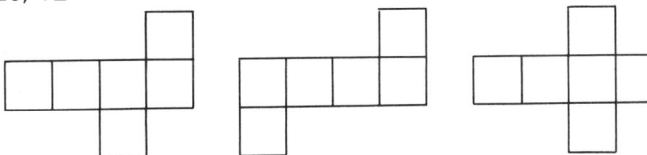

**Page 53**
1. a. 30°   b. 60°   c. 120°   d. 145° (approximately)   e. 30°   f. 60°   g. 90°
   h. 60°
2. a. 30°   b. 60°   c. 90°   d. 120°   e. 150°   f. 180°
3. a.                              b.

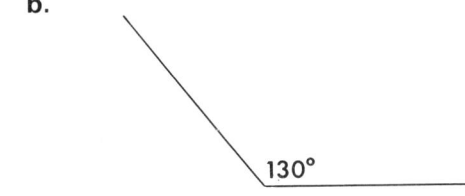

4. a.                              b.

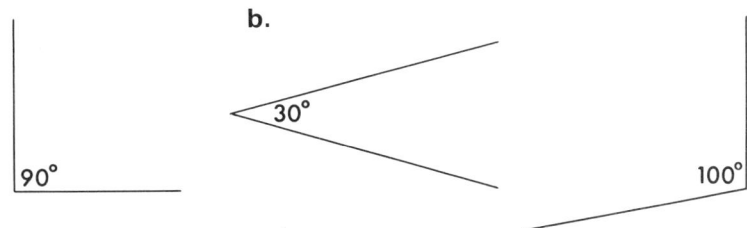

**Page 55**
1. a. 50°+130°+50°+130° = 360°   b. A and C   c. B and D
2. Sum in each case is 360°.
3. b. 90° each   c. 60°, 120°, 60°, 120°   d. No
4. c. 2 revolutions, or 8 right angles, or 720°

**Page 57**

**1. a.** An "Ian number"

**b.**

| + | I | A | B |
|---|---|---|---|
| I | I | A | B |
| A | A | B | I |
| B | B | I | A |

**2. a.**

| × | 0 | 1 | 2 | 3 |
|---|---|---|---|---|
| 0 | 0 | 0 | 0 | 0 |
| 1 | 0 | 1 | 2 | 3 |
| 2 | 0 | 2 | 0 | 2 |
| 3 | 0 | 3 | 2 | 1 |

**b.** No

**3. a.** Total in any row, column, or diagonal is 15.  **b.** Yes; total is 0 (mod. 5).
**c.** Yes; total is 3 (mod. 4).

**4. a.** Four

**b.**

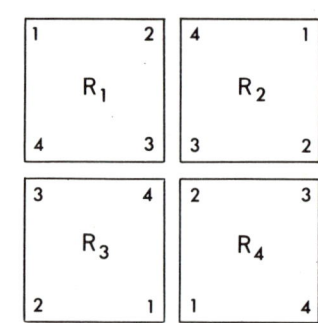

**c.**

| + | $R_1$ | $R_2$ | $R_3$ | $R_4$ |
|---|---|---|---|---|
| $R_1$ | $R_1$ | $R_2$ | $R_3$ | $R_4$ |
| $R_2$ | $R_2$ | $R_3$ | $R_4$ | $R_1$ |
| $R_3$ | $R_3$ | $R_4$ | $R_1$ | $R_2$ |
| $R_4$ | $R_4$ | $R_1$ | $R_2$ | $R_3$ |

**d.** Yes

**Page 59**

**1. a.** {(1, 1), (1, 2), (2, 1), (2, 2)}; {(1, 1), (1, 3), (3, 1), (3, 3)}
{(2, 1), (2, 2), (3, 1), (3, 2)}; {(1, 2), (1, 3), (2, 2), (2, 3)}
{(2, 2), (2, 3), (3, 2), (3, 3)}; {(2, 1), (1, 2), (3, 2), (2, 3)}

**b.** {(1, 1), (1, 2), (3, 1), (3, 2)}; {(1, 2), (1, 3), (3, 2), (3, 3)}
{(1, 1), (1, 3), (2, 1), (2, 3)}; {(2, 1), (2, 3), (3, 1), (3, 3)}

**c.** {(1, 1), (2, 2), (3, 2), (2, 1)}; {(2, 1), (1, 2), (2, 2), (3, 1)}
{(1, 2), (2, 3), (3, 3), (2, 2)}; {(2, 2), (1, 3), (2, 3), (3, 2)}
{(1, 1), (1, 2), (3, 3), (3, 2)}; {(3, 1), (3, 2), (1, 3), (1, 2)}

**c.** P = {(1, 1), (2, 2), (3, 3), (4, 4), (5, 5), (6, 6)}
Q = {(1, 6), (2, 5), (3, 4), (4, 3), (5, 2), (6, 1)}

**2. a.** and **b.**

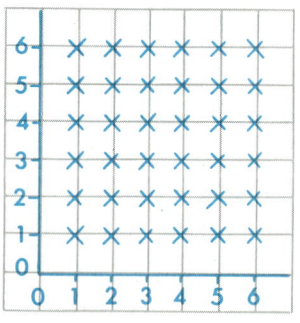

**d.**

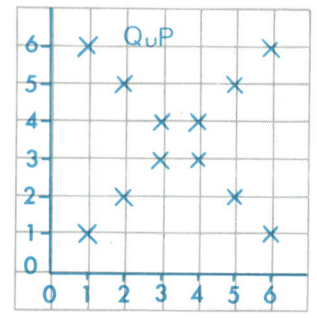

**3. a.** Yes  **b.** $R_1$ = {(A, 1), (B, 2), (C, 3)}
                   {(A, 1), (B, 3), (C, 2)}; {(B, 1), (A, 2), (C, 3)}
                   {(B, 1), (C, 2), (A, 3)}; {(C, 1), (A, 2), (B, 3)}
                   {(C, 1), (B, 2), (A, 3)}

   **c.** Six  **d.** Four

## Page 61
**1. b.** About 63 cm  **c.** About 3·1
**2. a.**

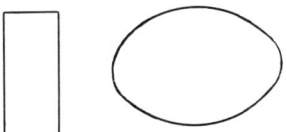

                         **c.** A circle

   **b.**                    Ellipse

**4. a.** 125·6 cm  **b.** 78·5 cm  **c.** 56·52 cm  **d.** 3·14 m or 314 cm

## Page 63
**1.** a, c, d
**2. a.** Two  **b.** One  **c.** Two  **d.** No
**3.**

**4. a.** 140 cm  **b.** 31·4 cm

## Page 65
**1.**

| Sleeping | Play | Eating | School | TV |
|---|---|---|---|---|
| 150° | 60° | 30° | 90° | 30° |
| $\frac{5}{12}$ | $\frac{1}{6}$ | $\frac{1}{12}$ | $\frac{1}{4}$ | $\frac{1}{12}$ |
| 10 | 4 | 2 | 6 | 2 |

**2.**

| Cars | Lorries | Vans | Scooters |
|---|---|---|---|
| 45 | 20 | 10 | 15 |
| $\frac{1}{2}$ | $\frac{2}{9}$ | $\frac{1}{9}$ | $\frac{1}{6}$ |
| 180° | 80° | 40° | 60° |

**3.**

| 10p | 5p | 2p | 1p |
|---|---|---|---|
| 108° | 54° | 108° | 90° |
| $\frac{3}{10}$ | $\frac{3}{20}$ | $\frac{3}{10}$ | $\frac{1}{4}$ |
| 30 | 15 | 30 | 25 |
| 3 | 3 | 15 | 25 |

**4.**  Light      72°
     Medium 204°
     Dark      84°

**Page 67**

**1.**

| A | 1 | 2 | 2 | 0 |
|---|---|---|---|---|
| B | 1 | 1 | 1 | 1 |
| C | 1 | 2 | 2 | 0 |
| D | 1 | 0 | 0 | 0 |

**3. c.** Slant height

**4.**

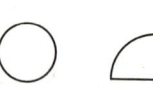

**5.** 2 curved faces, 2 edges, 1 vertex
**6.** Circular

**Page 69**

**1.**

| Network | Odd | Even | |
|---|---|---|---|
| A | 2 | 1 | Yes |
| B | 2 | 1 | Yes |
| C | 6 | 0 | No |
| D | 0 | 4 | Yes |
| E | 2 | 4 | Yes |

**Page 71**

**1. a.** 100000  **b.** 1000  **c.** 500  **d.** 100  **e.** 2000  **f.** 1000000
**2. a.** 1 800  **b.** 8  **c.** 1·2  **d.** 16·48  **e.** 12  **f.** 164·8  **g.** 1·8  **h.** 0·8
**3. a.** 94  **b.** 112·8  **c.** 124·8  **d.** 2·808  **e.** 4·998  **f.** 14·2252
**4. a.** 1 500 cm or 15 m  **b.** 625 cm or 6·25 m  **c.** 325 cm or 3·25 m
**5. a.** 375 m  **b.** 168·75 m  **c.** 16·875 m

**Page 73**

**1. a.** 2 km; 5 km; 5 km; 2 km
   **b.** 10.15 a.m. and 10.55 a.m.; 10.05 a.m. and 11.05 a.m.; about 10.27 a.m. and 10.47 a.m.; from 10.35 a.m. to 10.45 a.m.  **c.** 10 min  **d.** 12 km per hour; 4 km per hour
**2.** 10.40 a.m.

**3.** 98 min

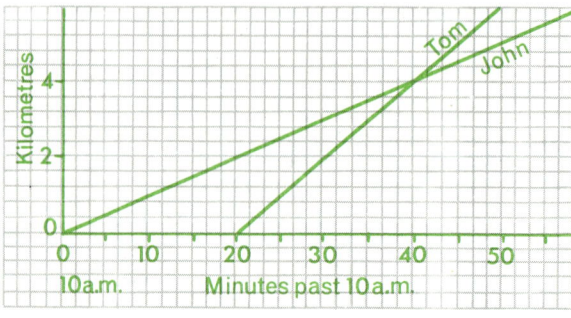

**Page 75**
1.  a. 1000   b. 196000   c. 250000
2.  a. 35 cm³   b. 70
3.  a. Yes   b. £31·50
4.  a. 1950 m³   b. 178 journeys (177$\frac{3}{11}$ loads)
5.  a. 1875 m³   b. 1875000 litres
6.  a. 78·75 m³   b. 60 crates   c. 8; 630 crates


**Page 77**
1.  a. 12·25   b. 12·425   c. 11·545   d. 28·2   e. 2·82   f. 14·887
2.  a. 0·125 m   b. 0·083 m   c. 12·5 cm, 8·3 cm
3.  a. 0·2 m   b. 20 cm   c. 60 cm
4.  a. 0·25 m   b. 25 cm
5.  a. 11·25 cm   b. 8·437 cm


**Page 79**
1.  a. 105   b. 320   c. 3000   d. 2
2.  a. 30   b. 45   c. 54   d. 258
3.  Area       a. 276 cm²    b. 216 cm²    c. 474 cm²
    Volume      280 cm³        216 cm³        630 cm³
4.  a. 19 m²   b. 1 m²   c. 10000 cm²   d. 10000
5.  a. 54   b. 8, 12, 6   c. One


**Page 81**
1.  a. 24 by 1, 12 by 2, 8 by 3, 6 by 4, 4 by 6, 3 by 8, 2 by 12, 1 by 24
    c. The point (16, 1$\frac{1}{2}$) should fall on the curve. Area = 24 cm².

2.  a.

| L | 1 | 2 | 4 | 6 | 8 | 10 | 12 | 14 | 16 | 18 | 19 | 20 |
|---|---|---|---|---|---|----|----|----|----|----|----|----|
| B | 19 | 18 | 16 | 14 | 12 | 10 | 8 | 6 | 4 | 2 | 1 | 0 |
| A | 19 | 36 | 64 | 84 | 96 | 100 | 96 | 84 | 64 | 36 | 19 | 0 |

    b. The square has the greatest area.


**Page 83**
1.  b. 20 books   @ 8p   £1·60     c. 25 pads   @ 7p   £1·75
    35 pencils @ 3p   1·05        50 pens   @ 6p   3·00
    24 rubbers @ 2p   0·48        60 rulers @ 4p   2·40
    36 rulers  @ 4p   1·44        45 books @ 8p   3·60
                      £4·57                       £10·75

    d. J. Smith: £1·35; J. Brown: £3
2.  a. £70+£200+£180 = £450   b. £55
3.  a. £24


**Page 85**
1.  a. Hours   b. Weeks   c. Days
4.  a. West   b. North

**Page 87**
1. **a.** 4  **b.** 4·2  **c.** 204  **d.** 16·562  **e.** 1·75  **f.** 66·6̇
2. **a.** 3·25  **b.** 4·054
3. **a.** 1·25 min  **b.** 9·6 m
4. **a.** 10  **b.** 5 (5·882)  **c.** 11 (11·764)  **d.** 23 (23·529)
5. **a.** 20  **b.** 68 (68·965)

**Page 89**
1. **a.** $\frac{4}{5}$  **b.** $\frac{2}{3}$  **c.** $\frac{3}{2}$
2. **a.** $\frac{5}{4}$  **b.** $\frac{3}{2}$  **c.** $\frac{2}{3}$
3. **a.** $\frac{5}{7}$  **b.** $\frac{4}{1}$
4. **a.** $\frac{2}{3}$  **b.** 7 m  **c.** 13·5 m  **d.** $\frac{1}{9}, \frac{1}{9}$

**Page 91**
1. **a.** A and E; B and F  **b.** B and F
4. **a.** Five  **b.** Five  **c.** AFGB, FEGA, EDGF, DCGE, CBGD

**Page 93**
1. **a.** 25 cm²  **b.** 25 cm²  **c.** 25 cm²  **d.** 55 cm²  **e.** 63 cm²  **f.** 64 cm²
2. **a.** 6 m²  **b.** 5 m²  **c.** 22 m²
3. **b.** 600 m²  **c.** 45°  **d.** About 108 m

**Page 95**
1. **a.** $^{-}2$  **b.** $^{+}2$  **c.** $^{-}5$  **d.** 7
2. **a.** $^{+}2$  **b.** $^{-}2$  **c.** $^{+}3$
3. **a.** (n+2, n)  **b.** (n, n+2)
4. **a.** (3, 0) and (4, 1)  **b.** (0, 3) and (1, 4)
5. **a.** (0, 2)  **b.** (2, 0)
6. **a.** 6  **b.** 2  **c.** 2  **d.** 1
7. **a.** 0  **b.** $^{+}3$
9. **a.** (3, 9)  **b.** (6, 6)  **c.** (3, 5)

**Page 97**
1. **a.** 104 g  **b.** 10 g  **c.** 56·25 g or $56\frac{1}{4}$ g  **d.** 45 g
2. **a.** A 40 cm³; B 20 cm³  **b.** $\frac{1}{24}$p, $\frac{1}{24}$p  **c.** 24 g, 24 g  **d.** $\frac{1}{2}, \frac{1}{2}, \frac{1}{2}$
   **e.** Both are same value.
3. Box A
4. **a.** 0·24p  **b.** 0·25p  **c.** No
5. **a.** 8·5 g  **b.** $44\frac{4}{9}$ g  **c.** $33\frac{1}{3}$ g

**Page 99**
1. **a.** A 1 in 3; B 2 in 1; C 2 in 1; D 3 in 1; E 1 in 1  **b.** D  **c.** A  **d.** B and C

2. **a.**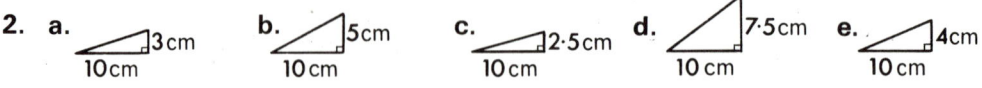

3.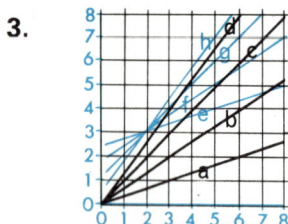

126

**Page 101**
1. $\frac{1}{5}$ = 20%, so 25% is the best reduction.
2. **a.** 6　**b.** 30　**c.** 60
3. **a.** 55%　**b.** 242 boys, 198 girls
4. **a.** 6 litres　**b.** 18 litres　**c.** £25·68
5. **a.** £15　**b.** £22·50　**c.** £24　**d.** £23·25

**Page 103**
1. **a.** 2　**b.** 6　**c.** 4　**d.** 14　**e.** 16　**f.** 16
2. 　　　　　　　　　　　　　　　　　　**d.** 7　**e.** 12　**f.** 11　**g.** 18

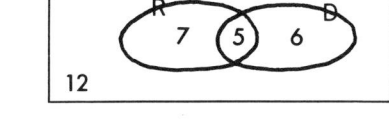

3. **a.** {1, 4, 9, 16}　**b.** {4, 6, 8, 9, 10, 12, 14, 15, 16, 18}

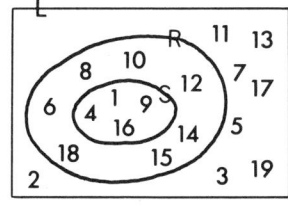 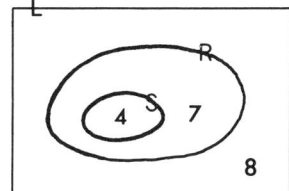

　　**e.** Eight numbers {2, 3, 5, 7, 11, 13, 17, 19}

**Page 105**
1. **c.** Four　**d.** Four　**e.** 90°, 180°, 270°, 360°
2. A **c.** Six　**d.** Six　**e.** 60°, 120°, 180°, 240°, 300°, 360°
　 B **c.** None　**d.** One　**e.** 360°
3. **b.** $R_{360}$　**c.** $R_{240}$　**d.** $R_{120}$　**e.** $R_{240}$

**Page 107**
1. **a.** 5 g　**b.** 1 g
2. **a.** Chalk absorbs water.　**b.** About 1 g per cm³

**Page 109**
3. **a.** $\overline{CD}$　**b.** $\overline{AB}$ and $\overline{CD}$　**c.** $\overline{DT}$
　**d.** About 52 metres plus 1·5 metres; total 53·5 metres

**Page 111**
1. she goes out; she out goes; goes she out; goes out she; out she goes; out goes she
2. 5! ways (5! = 120)
3. **a.** 5　**b.** 4　**c.** 3　**d.** 2　**e.** 1　**f.** 5×4×3×2×1 = 120
4. 1, 2, 6, 24, 120, 720, 5040, 40320
5. 7! ways

**Page 113**
3. Red $\frac{1}{2}$, green $\frac{1}{2}$; red is probably more popular.
4. **a. b. c. d. e.** $\frac{1}{6}$ each
6. **a.** $\frac{13}{20}$　**b.** $\frac{7}{20}$　**c.** 13 in 20　**d.** 7 in 20　**e.** 6 500　**f.** 3 500

127